Cape York, Greenland — p. 29

Siberia — p. 219

os Island, Bahamas —
p. 101

Barrier Reef — p. 113

JEROME J. NOTKIN

New Zealand — p. 229

Bringing the World Indoors

Landscapes, flora, fauna and minerals from all parts of the globe have been brought to the halls of the American Museum of Natural History. Displayed artistically, they give a graphic and enlightening picture of life on this planet.

Tierra del Fuego — p. 201

Little America — p. 277

THE WORLD OF
NATURAL HISTORY

THE WORLD OF NATURAL HISTORY

as revealed in

The American Museum of Natural History

by John Richard Saunders

ILLUSTRATED

SHERIDAN HOUSE · NEW YORK

Copyright 1952 by John R. Saunders

PUBLISHED SIMULTANEOUSLY IN CANADA
BY GEORGE J. MC LEOD LTD. TORONTO

PRINTED IN THE UNITED STATES OF AMERICA

To "Tib," *My Wife*

ACKNOWLEDGMENT

THE STORY behind any large and active museum is a story not merely of things but of people and their ideas. The American Museum of Natural History is now in its eighty-third year of growth and has become one of the largest and most outstanding institutions of its type in the world. Its complete story has never before been told. The author realizes that no single volume could ever contain the full story of the remarkable institution. He has tried, within the limitations of space, to select for the reader those items and incidents which in his opinion will give the reader the most complete picture of the American Museum of Natural History ever to have appeared in a single volume.

The purpose of this book is to give in pictures and in text the story of all of the museum's major exhibitions, its major expeditions and some of its significant researches; to give the reader a glimpse "behind the scenes" (an experience normally denied the museum visitor) and to highspot some of the exciting history of the growth of the institution.

Most of the material selected was chosen to enable those who have visited, or who are going to visit the museum, to obtain the background for enhancing their trip in the museum and to bring the museum to those who may not have the opportunity of coming to the museum.

The author wishes to express his sincere thanks to all those

ACKNOWLEDGMENT

who have aided in the production of this book, particularly to his wife, Agnes K. Saunders, who gave untiring editorial assistance, to Dr. Albert Eide Parr, Director of The American Museum of Natural History, who graciously granted the author permission to embark upon this extracurricular venture, to Mrs. Simone Sindin who typed the entire manuscript, and to his museum colleagues whose suggestions and encouragement lightened the task.

All photographs were obtained from the Photography Division of The American Museum of Natural History, except the following: The Gorilla Group by Wurts Bros.; Dr. T. C. Schneirla and his ant maze by Black Star and President Alexander White by Fabian Bachrach.

JOHN RICHARD SAUNDERS

TABLE OF CONTENTS

	Page
Young Man with an Idea	13
CHAPTER ONE: *Beyond This Planet*	29
CHAPTER TWO: *Beneath the Grass Roots*	51
CHAPTER THREE: *Probing the Past*	71
CHAPTER FOUR: *Animals Without Backbones*	96
CHAPTER FIVE: *Scales, Feathers and Fur*	100
CHAPTER SIX: *Man Before History*	168
CHAPTER SEVEN: *Native Americans*	187
CHAPTER EIGHT: *The Twain Shall Meet*	219
CHAPTER NINE: *Men of Science*	238
CHAPTER TEN: *Operation Expedition*	260
CHAPTER ELEVEN: *Behind the Scenes*	280
CHAPTER TWELVE: *Man and Nature*	295

FOREWORD

Young Man with an Idea

A LONE FIGURE walked slowly along the beach near Fort Macon in North Carolina. His gaze was constantly fixed on the sand at his feet. Occasionally he stooped over, picked up something, and placed it in a knapsack which hung loosely from one shoulder. Nearby, behind a hummock of beach grass, two figures in gray lay motionless; their rifles, which rested upon a slight ridge, trained upon the beachcomber in blue. Finally, one of the men in gray shifted his position and turned toward his companion. "We might as well go back. It's only him again, that crazy Yankee shell-hunter." The other Confederate grunted in disgust and drawled, "Third time this week we'uns hev had 'im in our sights. He all 'ill git a ball in his haid foah his troubles one day. Wondah what he wants with them ole clam shells anyhow? Le's go, Sargint, we is due back at foah an it's ten to now."

The lucky, "crazy shell-hunter" was a young infantryman named Albert Smith Bickmore. A nine-month soldier in the Union Army he was assigned to a Union hospital on the North Carolina coast. After arduous service at Tarboro and at Goldsboro in the Forty-fourth Regiment of Massachusetts Volunteers under Col. Francis L. Lee, Bickmore found his regiment inactive pending the reorganization of the Army of the Potomac after its stunning defeat at Fredericksburg. He had heard that the Medical Department at Washington had requested that a meteorological record be kept at as many of the army hospitals as possible. He applied for an assignment in this work and received one in North Carolina. When he found that his duties were not great and he had plenty of spare time, he wrote to his teacher at Harvard, the great naturalist Louis Agassiz, informed him of his situation and offered to spend his spare time collecting shells and marine specimens for Agassiz's Museum at Cambridge.

It was no accident, but rather the outcome of a single-minded devotion which had brought Bickmore to Harvard and Agassiz. As a boy, living in a house in Maine situated with especial good fortune between a forest and the sea, he had become preoccupied with the natural wonders of both woodland and ocean. His first ocean voyage was made in the barque of which his father was captain, a voyage which was the beginning of travels which further stimulated his interest in nature.

It was only natural, then, that he should have gravitated to Harvard for the very purpose of studying under Agassiz, the outstanding natural scientist of his day, and to be near the Agassiz Museum at Cambridge.

It was while he was a student that he conceived the great idea of his life—that of a great natural history museum, rivalling, if not surpassing, the best that Europe had to offer, and to be located in the city of New York, where millions could benefit by it.

In the partial, never published, autobiography which Bickmore wrote toward the end of his life, he tells with great enthusiasm how he broached his idea to the famous Dr. Ackland of Oxford University while the latter was on a visit to the Agassiz Museum, how Doctor Ackland's approval encouraged him to draw plans for the building which was to house his museum, how later Dr. Ackland in London introduced Bickmore to Sir Richard Owen, superintendent of the Natural History department of the British Museum, who showed the young American the plans for that great structure—a central lecture hall from which radiated groups of buildings. His own ideas modified and improved, and his courage and enthusiasm heightened by the approval of the two famous Englishmen, Bickmore came home and began seriously to solicit support for his plan.

Two other factors meanwhile played into his hands.

In 1836 the State of New York undertook a survey of its natural history which, under the direction of Professor James Hall, was to continue for over fifty years. During its course many valuable natural history specimens were gathered and housed in a temporary museum at Albany.

Another factor of great importance was the opening of Central Park on Manhattan Island in the city of New York. The new park, among other things, was intended to afford a site for buildings of public interest, such as museums and the homes of scientific societies.

New York was beginning to lag behind other cities. Already Philadelphia, Boston, Washington, and Chicago had taken steps toward getting permanent homes for their natural history collections. The Academy of Natural Sciences in Philadelphia, which had been organized in 1812, had secured a building for its collections in 1840. In Washington, D.C., the Smithsonian Institution had been in existence a quarter of a century. Agassiz had already succeeded in getting a museum at Cambridge, and in 1864, under the auspices of its Society of Natural History, Boston obtained its museum. The Chicago Museum, already established, had recently been seriously damaged by fire.

In New York, however, no concerted action to produce a natural history museum had been taken. There were several collections, some of little importance, some of popular character such as P. T. Barnum's American Museum, and a few, unrelated and independent, of scientific calibre. The most important of these, the New York Academy of Science, which had accumulated valuable collections, had never had a permanent home for them, and they had been destroyed by fire in 1866, when housed in the Medical College. Thus the time and the circumstances were right for Bickmore's plan when he returned from abroad.

He went at once to the home of his friend and benefactor, Mr. William E. Dodge, who had financed his trip, and who was interested in the plan for the new Museum. However, Dodge was extremely busy with his own responsibilities at the time and he sent Bickmore to see Mr. Theodore Roosevelt, father of the famous "Teddy" Roosevelt. This meeting was to prove the beginning of a long and valuable association between the Roosevelt family and the Museum. Years later, "Teddy" Roosevelt was to go on an expedition to South America with George Cherrie of the Museum and was to donate many fine specimens of wild life to the Museum. The Museum was also to be the site of the Theodore Roosevelt Memorial Building, a New York State Memorial to its famous son who had been its Governor and later the President of the United States of America. When the Theodore Roosevelt Memorial Building was dedicated, President Franklin Delano Roosevelt related to the assembled audience the story of how his grandfather gave him a life membership in the Museum and told him, "Franklin, you can learn more about nature in this Museum than you can in all of the books of the world."

Theodore Roosevelt was the right man for young Bickmore to see. With his help a letter was drafted to Andrew H. Green, commissioner of the newly established Central Park expressing the desire of "a number of gentlemen . . . that a great Museum of Natural History . . . be established in Central Park." Bickmore took the letter around New York asking for signatures, and on December 30, 1868, the letter went off, signed by James Brown, Alex. T. Stewart, Benj. H. Field, Adrian Iselin, Robert L. Stuart, Marshall O. Roberts, Theodore Roosevelt, George Bliss, Morris K. Jesup, William T. Blodgett, John David Wolfe, Robert Colgate, I. N. Phelps, Levi P. Morton, William A. Haines, J. Pierpont Morgan, A. G. Phelps Dodge, D. Jackson Steward, and Howard Potter. Two weeks later a reply was received from Green expressing the complete approval of the Central Park commissioners.

With this encouragement, the nineteen signers of the original letter met in Benjamin Field's home and passed resolutions to the effect that they would take steps to purchase the Verreaux Collection, then up for sale in Paris, and that a

Committee be appointed to raise the necessary funds and to act as Trustees for the dispensing of these funds. New names appeared in this first group of trustees. In addition to the signers of the letter of application to Green, the names of Benjamin B. Sherman, Charles A. Dana, Joseph H. Choate and Henry Parish appeared.

Joseph H. Choate was asked to draw up a charter. Bickmore described to Choate the organization of the Agassiz Museum and told of its relationship to Harvard University and the educational system of Massachusetts. From the very beginning there were indications that the support of the new museum was not to be entirely private, that the City would have its share of responsibilities along with the Trustees. Because the Trustees wanted a name which would be national in character, and one which would express the hope that their museum would ultimately become the leading institution of its kind in the country, the new venture was called The American Museum of Natural History.

Choate's draft of a proposed charter was unanimously approved by his fellow founders and Bickmore, commissioned to see that the act of incorporation was passed in Albany, obtained the help of State Senator William Tweed, the man known to political historians as "Boss Tweed." On April 9th, 1869, with Governor Hoffman's signature to the incorporating act, the American Museum of Natural History became a reality in the eyes of the law. It still was a museum on paper only and much remained to be done before it became a physical reality.

The early relationships of the Museum with the City of New York set a most effective pattern for museum-city cooperation. It was arranged that the Museum Corporation was to furnish the collections and was to provide the scientific and educational program while the City of New York was to furnish the land and buildings and provide for the maintenance of the physical plant. This was a new idea in municipal enterprise which has since been incorporated into the contract between the Metropolitan Museum of Art and the City of New York, and still later into the contracts between the City and the New York Zoological Park and the Botanical Gardens. It is a wise and an excellent adjustment in both theory and practice, superior in many ways to exclusive municipal, state, or private control.

With John David Wolfe as the Museum's first President,[1] and Albert Smith Bickmore as its first Director (though he was known as Superintendent, in accordance with the custom of the British Museum), money began to come to the new institution. By November of the first year some $44,500 had been subscribed.[2] But the Museum was as yet without a home. In the beginning it had nothing to house, to be sure, but it was in the process of negotiating for the purchase of its first collections and the need for storage space was urgent. The Trustees appealed first to Cooper Union, that "hospitable center for all public efforts in education and culture," and then to the Commissioners of Central Park for temporary quarters, and the latter made available the second and third floors of the old Arsenal Building on the Fifth Avenue side of the park near 64th Street. This building, now serving as the office for the Park Department, is

[1] The Museum has had but six presidents throughout its life: John David Wolfe, 1869-72; Robert L. Stuart, 1872-81; Morris K. Jesup, 1881-1908; Henry Fairfield Osborn, 1908-33; F. Trubee Davison, 1933-51, and Alexander White, 1951—.

[2] Today it costs over $2,000,000 to maintain the Museum and its work for a year.

an oblong structure four stories high whose architecture still betrays its original use as an Arsenal. As a museum building it was picturesque but most inadequate. Cases were furnished by the Park Commissioners, repairs were made and an ample application of paint and putty made the Arsenal home of the Museum as presentable as possible. Happily it was right next to the park menagerie, which had already become a favorite haunt for New Yorkers and out-of-town visitors and many of these found their way from the living creatures in cages to those which were dead and in cases. The first exhibits bore little resemblance to the masterpieces of creative exhibition one finds in the Museum today. They resembled more closely what we now call study collections, case after case simply filled with specimen after specimen, arranged according to families or according to geographic distribution, and without added material. In most museums today such arrangements of specimens in a highly systematic way are not featured as public exhibits. They are often made available only to seriously interested students whose studies demand the examination of such material.

The Trustees of the new Museum set out to acquire the best collections their limited means would permit. Among the first of the Museum's collections were these four: a collection of some 2,500 birds gathered by the ornithologist D. G. Elliot; the important collections of Prince Maximilian of Neuwied on the Rhine above Bonn, comprising 4,000 mounted birds, 600 mounted mammals, and about 2,000 fishes and reptiles mounted and in alcohol; and the principal parts of the Vedray and the Verreaux collections.

The Verreaux Collection was considered one of the finest in the world at the time. As early as Nov. 13, 1868, Bickmore wrote to W. E. Dodge regarding the possible purchase of this collection, before the Museum had been incorporated, pointing out that it contained "about 1,300 specimens, which is more than the Smithsonian Institution, the Academy at Philadelphia, the Society of Natural History in Boston and Professor Agassiz's Museum together have on exhibition. . . . I am hoping providence will open some way for its coming to Central Park, for in this way we would have secured a Museum in a day!" Actually the Verreaux Collection as purchased contained 2,800 mounted birds, 220 mounted mammals, and about 4,000 mounted skeletons of mammals, birds, reptiles, and fishes.

The Vedray Collection consisted of 250 specimens of mounted mammals and birds of Siberia. Today, the Museum's mammal collections contain upwards of 150,000 specimens, its bird collections number almost 1,000,000 specimens, its anthropological specimens total some 2,000,000 items, and its other vast collections of fossils, insects, amphibians and reptiles, invertebrates, rocks and minerals, meteorites and fishes, add up to more millions.

As additions were constantly added to the four original collections, the need for a permanent museum building became more and more evident, for the arsenal quarters were both unsuitable and unsafe. On Nov. 14, 1870, trustees Blodgett, Roosevelt, and Morgan were appointed "to take in charge and procure such legislation as may be necessary to promote the interests of the Museum," the "interests of the Museum" in this instance meaning a permanent home. The Metropolitan Museum of Art, truly a sister institution, with some of its personnel partly identified with the American Museum of Natural History, was in even a worse position. It had no building at all, not even a temporary one, its collections moving about from one private residence to another. The Metropolitan joined forces with the American Museum and the joint efforts of both institutions resulted in the preparation of a great petition, signed by forty thousand New Yorkers, which asked for an appropriation of funds with which to begin the building of two museums for New York, one of Art, and one of Science. In response to this strong appeal the Legislature passed an act which authorized the Department of Parks to "contract for, erect, and maintain in and upon that portion of Central Park, formerly known as Manhattan Square, or any other public park, square, or place in New York City, a suitable fireproof building for the purpose of establishing and maintaining therein a museum of natural history, to be occupied by the American Museum of Natural History."

The exact permanent location of the American Museum was not decided quickly. Some of the authorities felt that the new Museum should be located near the Menagerie, or in other words, near the Arsenal. Actually the first site offered the Museum by the City was the present site of the Metropolitan Museum of Art on the Fifth Avenue side of Central Park, near 83rd Street. Another site that was given careful consideration was that which was known as Reservoir Square, the present site of the New York Public Library, running from 41st to 42nd Streets between Fifth Avenue and the Avenue of the Americas (Sixth Avenue). These early offers were rejected by a special trustee committee and the final selection was the present Manhattan Square site.

Manhattan Square is a plot of land, some 18 acres in extent, running from 77th Street on the south to 81st Street on the north and from Central Park West (8th Avenue) on the east to Columbus Avenue (9th Avenue) on the west. Today it is amply served by subway and bus, with a direct entrance from the 81st Street Station of the Eighth Avenue Subway into the basement floor of the Museum. Beautiful Central Park adjoins the Museum grounds on the east and that section of Manhattan Square still unoccupied by museum buildings is landscaped and sprinkled with shrubs, shade trees and benches. In 1869, however, Central Park was still a rugged tract of land, for the most part a park in name only. Sections, including Manhattan Square, consisted of shanty-crowned rocks over which goats wandered, and squalid hollows where the squatters, or inhabitants of shantytown, were wont to raise their ragged patches of vegetables. A small pond occupied one of the hollows of Manhattan Square and although the site of the original pond is now covered by museum buildings, the spring which fed the pond is still active and despite efforts to control it, it occasionally causes basement floods during rainy spells. In fact, the Hayden Planetarium at the Museum had to be supported on piles of concrete driven eighty feet into the unstable spring bed.

In 1869 only a single horse-car line ran along Eighth Avenue, close to the edge of the pond in Manhattan Square. The bulk of the city's population still lived south of the new Museum's site, which now, thanks to New York's rapid development, is actually centrally located.

Once the opportunity for getting a permanent site and a permanent building presented itself, the Trustees resolved that the architects R. M. Hart, Russell Sturgis Jr. and Renwick and Sands be invited to unite in preparing designs for the first building. Bickmore, it will be remembered, had long since prepared a skeleton plan for the Museum—a simple plan calling for a T-shaped building, a design based upon the form of a single exhibition case which Bickmore had seen and liked because of its adaptability. (Such T-shaped cases were used in the old Birds of the World Hall which stood for many years on the second floor of Section One, the original building on the Manhattan Square site. A series of tall, slit-shaped, ground glass windows in the outer wall of the building, opened into the ends of the adjoining cases. Thus the problem of interior lighting of exhibition cases was solved before the day of electric lights.) Bickmore's plan proposed that each hall be sixty feet wide and one hundred and thirty feet long. In the plans that were finally adopted, the long halls were made sixty feet wide and the length was increased to one hundred and forty feet.

Calvert Vaux, Architect of the Park Department, was asked to be the architect for the new museum. His accepted design was for a building, or group of attached buildings which would form a great hollow square the sides of which were to be 740 feet long. There was to be erected in the center of the square thus formed, a tall structure terminating in a tower 400 feet high. From this center tower wings would radiate to the center of each of the four long buildings forming the sides of the square. These radiating wings would divide the center square into four large open courtyards which would serve to provide light to the interior sides of the exhibition halls.

For some time building proceeded according to the original plans of Vaux, but in later years the plan has been modified in the light of developments in modern architecture and in Museum construction.[1]

The early trustees of the Museum selected for the first unit to be constructed one of the interior wings, that which radiates from the center to the middle of the long south facade. There was a specific reason for their plan. The first building which the Museum would present to the public would have an unfinished look on the outside. To get to its door people would have to walk along a temporary wooden trestle-like structure. This, the trustees thought, would show the public that the Museum was meant to occupy the whole of Manhattan Square and would serve to remind the city officials that this first section was but the beginning. The idea seems to have paid off. The early construction which followed this first section appeared to be an effort to hide from public view this first, very unfinished-looking wing by shielding it with the more finished appearing outside buildings.

The cornerstone of the first section was laid with great ceremony on June 2, 1874. It was a gala occasion for New York. The President of the United States of America, Ulysses S. Grant, and three members of his Cabinet were present.

---

[1] Today, the long building forming the south side of the square is complete, that forming the east side is almost complete and the one on the west side is about one third complete. The base of the center building is complete and two of the wings radiating to the east and to the south from the center are finished. It is unlikely that the great central tower will ever be built. Since the advent of electricity, the open interior courtyards are no longer a necessity and the Museum has been adding buildings in these old courtyard areas. Four structures now occupy the four proposed courtyards. They are the Hayden Planetarium, the Ocean Life Building, the Power Plant and Preparation Building and the School Service Building. The building as it now stands is the largest museum in the United States and one of the three largest in the World. According to the original Vaux plans it is somewhat less than half-finished.

Governor Dix of New York State, the Mayor of the City, and many other dignitaries of the day attended. One of the principal addresses was delivered by Professor Joseph Henry, Secretary of the Smithsonian Institution. Henry prophetically stated that fine as the Museum was to be, it would still be finer should it actively engage in teaching its wonders to the people. The varied and wide-reaching activities of the present Museum's Education Department would have astounded and gratified Professor Henry. Henry also alluded to the new Museum as a possible "college of discoverers," a group of men "capable not only of expounding established truths but of interrogating nature and of discovering new facts, new phenomena, and new principles." The important contribution of the Museum's men of science in laboratory, field and classroom have gone a long way toward fulfilling Joseph Henry's wish.

Work continued on the first section. Meanwhile the small staff, mainly volunteers, were busy refurbishing the collections in the Arsenal and setting up new collections which were rapidly beginning to find their way to the young institution.

One of the early prizes of the Museum's treasures was a specimen of the extinct Great Auk. In 1871 Trustee Robert L. Stuart, later to become the Museum's second President, had presented this rare bird to the Museum. The public in those days knew little of nature. They thought of museum exhibits as containing the rare, the freakish, the unnatural. They had been conditioned by museums like Barnum's, where cherry-colored cats, mermaids and the like were the usual order of the day. A farmer from up-state made the long trip to the new Museum just to see the Great Auk. Bickmore conducted the eager visitor to the case containing the bird. The farmer took a long look, an expression of disappointment began to creep over his face, and as he turned to leave the exhibit, he was heard to murmur, "I've got bigger geese than that on my own pond."

In 1873, the building up of the study collections, as distinguished from the exhibition collections, began. From that time to the present, the Museum has gained increased importance because of the quality and significance of its collections and of the research based in no small degree upon these treasures.

Louis Agassiz once remarked, "Whoever gets Hall's collection gets *the* Geological Collection of America." Professor James Hall of Albany, N.Y. had amassed the largest and most complete collection of invertebrate fossils which had ever been made in America. While most of the specimens came from New York State and were gathered as part of the great survey of the natural history of New York which Hall headed there were also numerous specimens from the West. The Trustees acquired this prize collection for $65,000.00, no small sum in those days, especially for such a young institution.

The Museum got more than specimens when it got the Hall collection. It acquired the services of Professor R. P. Whitfield as its first Curator, the Curator of Geology. Whitfield had been Hall's assistant and he served the American Museum for over thirty years.

To tell the entire history of such a complex institution as The American Museum of Natural History would require several volumes and would hardly allow for the story of the World of Natural History contained under its roof. It should however serve as an inspiration to Americans that such a tremendous and beneficial institution could have come into being as the cooperative effort of its planners, its loyal friends who have contributed valuable counsel and financial means, the foresighted and cooperative government officials and the host of taxpayers whose monies went to support the work.

The World of Nature which the American Museum of Natural History brings to the American people is the product not of one mind but of many, but the catylist which served to bring it into reality was a young man with an idea, a boy who lived between a forest and the ocean, a Union soldier who was so obsessed with the task of studying nature that he strolled into the Confederate lines searching for shells for his teacher and inspiration, Louis Agassiz.

CHAPTER ONE

Beyond This Planet

NEW YORK has seen many strange spectacles on its streets but it saw one of the strangest on the day when a "star" was hauled up Broadway by a team of thirty horses. The "star" was the famous "Ahnighito," thirty-six and one half tons of black meteorite, which Admiral R. E. Peary had discovered and brought back from Cape York in Greenland in 1897. This solid chunk of nickel and iron, the world's largest known meteorite, proved to be most difficult to ship. To begin with, the Eskimos who helped Peary load the giant on his boat were quite unhappy about the whole job. They were superstitious people and believed that it was courting bad luck to move the big, black stone they called Ahnighito, or "The Tent." As the meteorite was being worked up to the side of the ship on skids, and the ship listed badly, the Eskimos felt certain that the bad luck attendent to the moving of the

Ahnighito would result in a capsized vessel. It is a tribute to the engineering skill of Peary and his men that the great object was finally moved from Cape York to the American Museum of Natural History without serious mishap.

At first the Ahnighito was placed on exhibit in the 77th Street Foyer of the Museum along with some of the other large meteorites in the Museum's outstanding collection. When the Hayden Planetarium opened in 1935, it and its companions were moved again, but this time the journey only involved going halfway around Manhattan Square.

The Ahnighito is the showpiece of the Museum's meteorite collection which is now entirely housed in the Planetarium, one of the three finest collections of meteorites in the world, and the third largest in the number of falls represented. These thousands of visitors from beyond our planet range in size from small grains resembling fine sand to large chunks of solid matter weighing tons.

The "Star" that Peary brought back from Greenland is, of course, not a star at all. People have called meteors which they see flashing across the sky, "Shooting Stars" but compared with real stars they are but tiny bits of matter. Stars are suns. Earth's beneficent, but awesome Sun, which is 864,000 miles in diameter is but one of the smaller stars. One is then thankful indeed that the Earth has never been hit by a real star. These lifeless, cold bits of cosmic debris which we call meteorites *after they land on the earth,* may be the remains of comets which have disintegrated. When they enter the Earth's atmosphere they are heated by friction caused by the tremendous speeds at which they travel, speeds ranging from six to almost fifty miles per second. A bullet shot from a rifle travels about a half-mile per second.

BEYOND THIS PLANET

Billions of meteors, most of them no larger than a tiny grain of sand, enter our atmosphere every day. Most of them burn and fall to the Earth as dust, but occasionally a larger one comes through the atmosphere without being consumed and actually strikes the Earth. There are records of over 1,000 such falls and specimens from about 550 of these are represented in the Museum's collection.

There is a fascination about meteorites. Actually, they are, outside of the Earth itself, the only other heavenly bodies we can touch. There are two major types or classes. Iron meteorites are composed of about nine-tenths iron, and nickel, cobalt, silicon, sulphur or other elements comprise the remainder. Stone meteorites after passing through the Earth's atmosphere are covered with a thin coating which is dark brown or black, but when analyzed these show little or no metal content. The Museum's collection has excellent examples of both types.

Peary also brought back from Greenland two smaller iron meteorites found near the Ahnighito and called by the Eskimos "The Woman" and the "Dog," "The Woman" weighing 6,000 pounds and the "Dog" 1,100 pounds.

The second largest meteorite in the collection is the Willamette. This specimen, about half the weight of "The Tent" looks almost as large as it is pock-marked with large depressions and holes, almost like a Swiss cheese. The Willamette came from near Portland, Oregon and is the largest of the American meteorites on exhibition. It weighs 15½ tons and was discovered in 1902. In 1906 Mrs. William E. Dodge bought it for $20,000 and gave it to the Museum.

The Museum's Department of Astronomy is of fairly recent origin. The meteorites, actually the only tangible astronomical specimens which can ever be exhibited in any museum until the time comes when inter-planetary travel becomes a reality, are under the charge of the Geology Department. For many years there was a small astronomical exhibition in addition to the meteorites. It consisted of some paintings of solar phenomena, hydrogen prominences on the sun, solar and lunar eclipses, scenes of the planets and some close-up paintings of the moon's rough surface. There also was a mechanical planetarium of the Copernican type wherein mechanically operated metal spheres, roughly approximating in scale the members of the solar family, revolved about a central lighted globe representing the sun. Such devices, very effective in demonstrating certain astronomical principles, are still to be seen in many museums. The Copernican Planetarium on the first floor of the Hayden Planetarium is a device of this type.

Plans for a separate astronomical section of the Museum were first developed in 1925 by the astronomer-artist, the late Howard Russell Butler, in collaboration with the Museum staff and the architects, Trowbridge and Livingston. Soon after the first Zeiss Projection Planetarium had been invented by Dr. Walter Bauersfeld, a research engineer of the Carl Zeiss Optical Works at Jena, Germany, Dr. George Clyde Fisher, first curator of the Astronomy Department, and a member of the staff of the Education Department, went abroad to examine the marvellous instrument. Here was something new under the Sun, of obvious importance as a visual aid in education.

Unlike the older type of Copernican Planetarium, the new instrument does not depend upon globes to represent the planets and the Sun and Moon, but rather upon the projection of light, a new principle in the construction of planetariums. Now not only the Sun and the Moon and the planets can be shown, but comets, meteor showers, the northern lights, and many other astronomical displays, can be depicted with an uncanny realism. What is even more important the starry heavens can be reproduced and all of the stars which one can see from the earth with the naked eye shown. The major feature of the invention is that all the real and apparent motions of the heavenly bodies can be simulated and this is done so accurately that the machine can be set to show the arrangement of the Heavens on any date from approximately 13,000 years ago to about 13,000 years in the future.

Two American instruments of the past might be considered as forerunners to the projection planetarium. In 1758 Professor Roger Long built a celestial sphere at Pembroke College, and in 1913 another was built by Dr. Wallace Atwood for the Chicago Academy of Sciences Museum. These devices are hollow globes inside of which the audience sits on a fixed platform. The globe is then rotated to simulate the apparent motion of the heavens. Bright stars are represented by little holes through which light passes from the outside. These celestial globes may be considered as intermediate between the Copernican type of planetarium with spheres to represent planets and the new projection type of planetarium.

The Deutches Museum in Munich was the site of the first Zeiss Projection Planetarium which was opened to the public in May, 1925. It was not until five years later that the first mechanism of this design was installed in America. This was the famous Adler Planetarium in Chicago which was made possible through the generosity of Max Adler. In 1933 the second machine, named after its donor, Mr. Samuel Fels, went into operation in Philadelphia at the Franklin Institute. Griffith Park in Los Angeles received the third Zeiss projector to come to America in May of 1935. New York's Hayden Planetarium was opened to the public in October 1935, the fourth of its kind in the United States. Today there are over 25 of these instruments scattered throughout the world and no better device for teaching the wonders of the heavens has been devised by man.

At the Museum's Hayden Planetarium, under the direction of Curator Robert R. Coles, a skilled staff of lecturers and technicians operate the projector and devise the many fascinating shows that not only demonstrate to the visitors the marvelous abilities of the instrument but create an almost unbelievable illusion of reality. The instrument, housed in an $800,000 building, was donated by Mr. Charles Hayden, New York financier, who gave $150,000 to provide for its purchase and installation. The projector is installed in a circular room, with a high domed ceiling. The diameter of the hemispherical dome on which the images are projected is 75 feet. The room seats approximately 750 persons making it one of the largest of the planetariums. On the floor in the center of the room stands the projector itself, a weird-looking instrument shaped somewhat like a giant dumb-bell. It is twelve feet long and at each end is a large globe. These globes are the projectors for the so-called "fixed stars." Of course all stars are in motion but they are so far away from the earth that though they are traveling at tremendous speeds, they appear to be fixed, and the heavens change but

Albert Smith Bickmore, the "father" of The American Museum of Natural History. Bickmore conceived the idea of the great natural history museum while a student of Louis Agassiz at Harvard University.

Henry Fairfield Osborn, founder and first curator of The American Museum's department of Vertebrate Paleontology. Osborn later served (1908 - 1933) as the fourth President of the Museum's board of trustees.

Impressive entrance hall of the Theodore Roosevelt Memorial, erected by the people of the State of New York. "Teddy, the Rough Rider" had a lifelong interest in natural history; his father was a member of the original board of trustees of the Museum. The walls of this spacious hall are adorned with mural paintings dealing with the varied career of Theodore Roosevelt and with quotations from his writings.

Above: The Theodore Roosevelt Memorial faces New York City's famous Central Park. Four Ionic columns fifty-four feet high are topped by statues of Boone, Audubon, Clark and Lewis, pioneers in the early exploration of The United States of America.

Opposite: The heroic equestrian bronze of Theodore Roosevelt by John Frazer stands in front of the Roosevelt Memorial.

Six presidents have guided The American Museum's destinies since 1869; John David Wolfe (1869 - 1872), Robert L. Stuart (1872 - 1881), Morris Ketchum Jesup (1881 - 1908), Henry Fairfield Osborn (1908 - 1933 [see portrait on previous page], F. Trubee Davison (1933 - 1951), and Alexander White (1951 -).

JOHN DAVID WOLFE

ROBERT L. STUART

Morris Ketchum Jesup    F. Trubee Davison    Alexander White

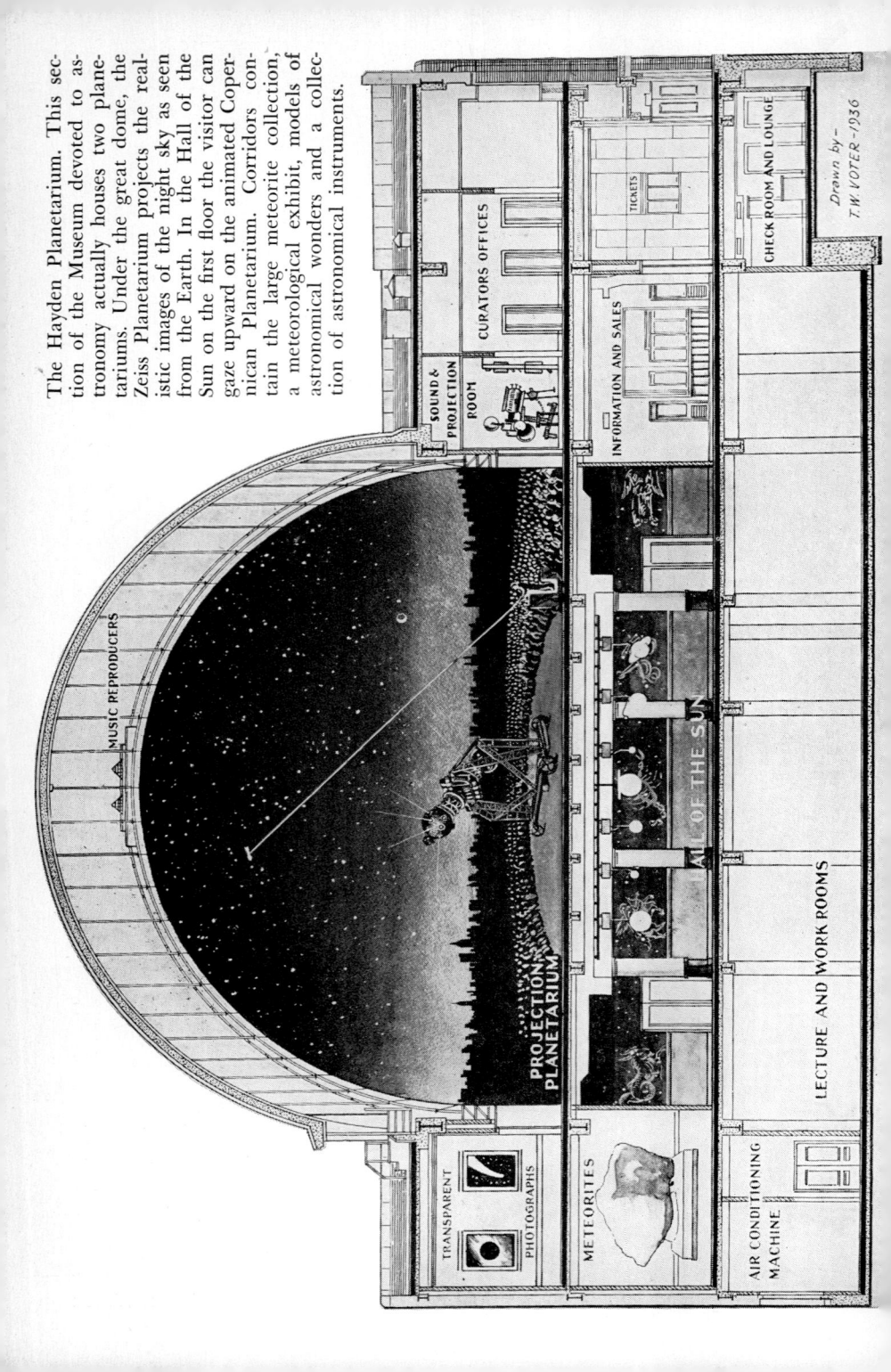

The Hayden Planetarium. This section of the Museum devoted to astronomy actually houses two planetariums. Under the great dome, the Zeiss Planetarium projects the realistic images of the night sky as seen from the Earth. In the Hall of the Sun on the first floor the visitor can gaze upward on the animated Copernican Planetarium. Corridors contain the large meteorite collection, a meteorological exhibit, models of astronomical wonders and a collection of astronomical instruments.

Drawn by—
T. W. VOTER – 1936

Above: G. Clyde Fisher, first director of the Hayden Planetarium stands beside the 36 ton *Ahnighito,* largest meteorite in any museum in the world.

Below: The "Star of India" is the largest cut star sapphire in the world. It weighs 563 carets and is one of the outstanding gems in the Museum's Morgan Hall.

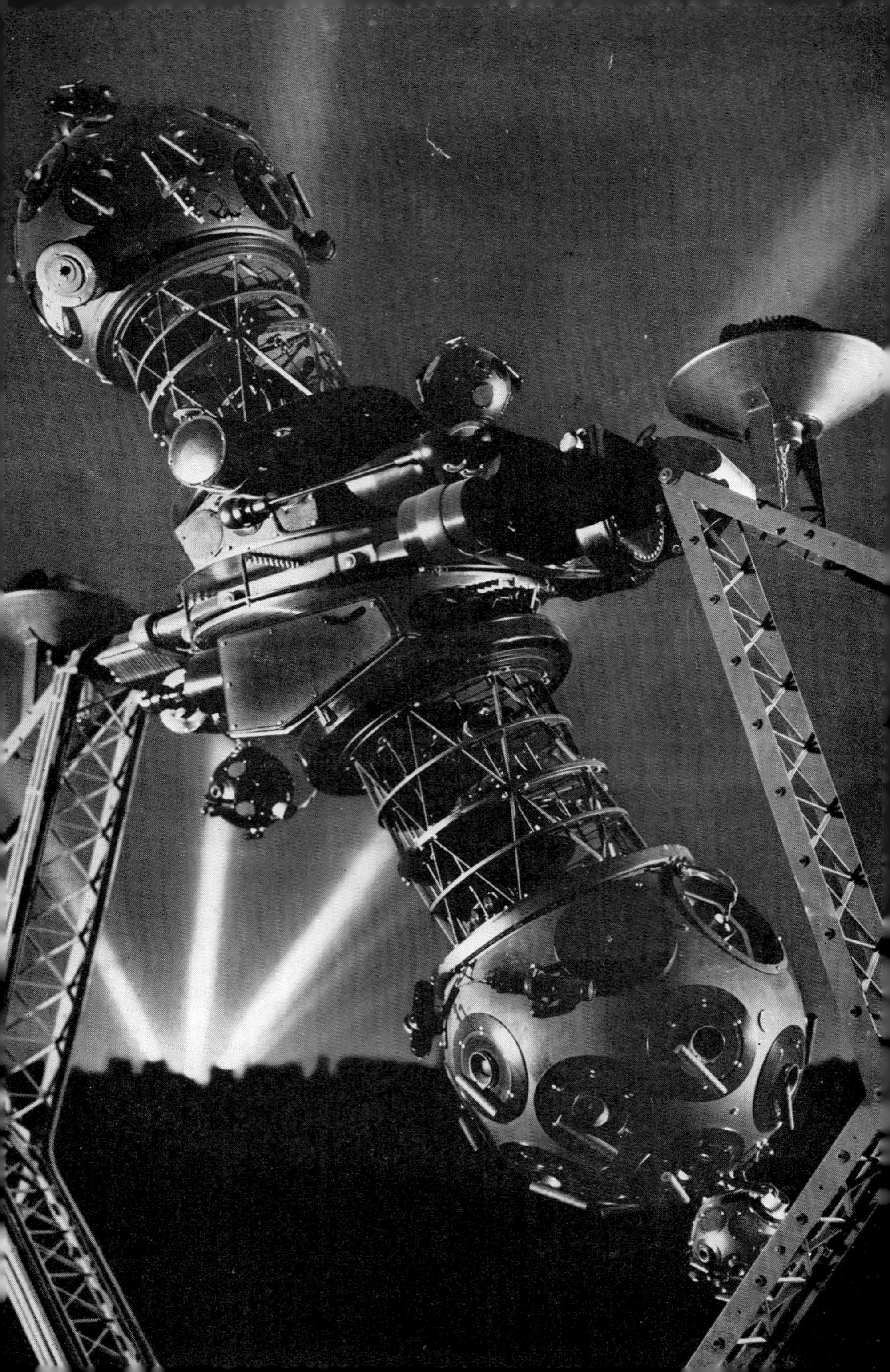

Left: Close-up of the Zeiss Projector at the Hayden Planetarium. This weird-looking instrument can reproduce all of the "naked-eye" stars, the sun, moon and planets in their apparent and real motions. *Above:* A scene from "The End of The World" show. *Below:* The Planetarium houses this exact duplicate of the Westinghouse Time Capsule which was buried in a concrete shaft on the site of New York's World's Fair, Flushing Meadows, 1939.

Above: Brontosaurus, a huge herbivorous dinosaur that lived more than 120,000,000 years ago. Some of these creatures reached a length of over 80 feet. A painting by Charles Knight shows some of these "Thunder Lizards" in a prehistoric marsh.

Below: Eohippus, four-toed ancestor of the modern horse, lived in the Lower Eocene almost 60,000,000 years ago, and was no larger than a small dog. Restoration by Charles Knight in the Hall of Fossil Mammals.

Above: Excavating bones of the giant *Baluchitherium* in the Gobi Desert of Mongolia. This prehistoric mammal towered 20 feet high at the shoulders and was probably the largest land mammal of all time.

Below: Restoration showing the small dinosaur, *Protoceratops andrewsi* and a nest of its eggs. The actual eggs in the Museum collection were discovered in Mongolia by the Central Asiatic Expedition.

The "King of the Tyrant Lizards," *Tyrannosaurus rex*, the largest carnivorous land creature the world has ever known. The Museum's mounted skeleton of this terrible engine of destruction is 47 feet long and its head towers 18½ feet above the ground. It lived in the Cretaceous Period about 100,000,000 years ago, probably weighed six to eight tons. Its small front legs were used only for grasping.

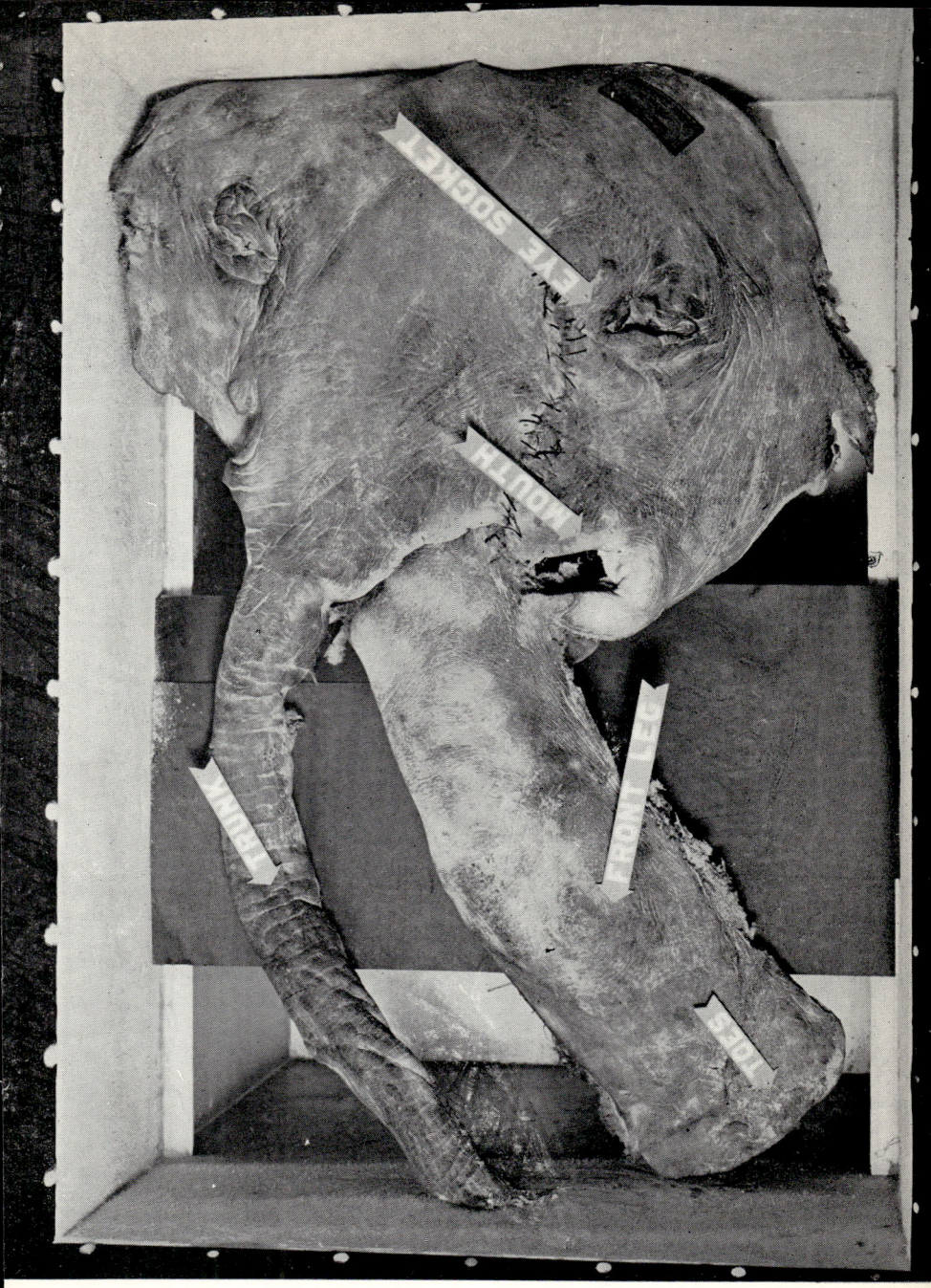

The partial remains of a baby mammoth now preserved in a deep-freeze unit at The American Museum. This frozen carcass was found in the tundra of Alaska and was the first prehistoric animal to be exhibited "in the flesh" at the American Museum. The remains consist of the head, trunk and one front leg. Creatures like this lived in North America in late glacial times more than 25,000 years ago.

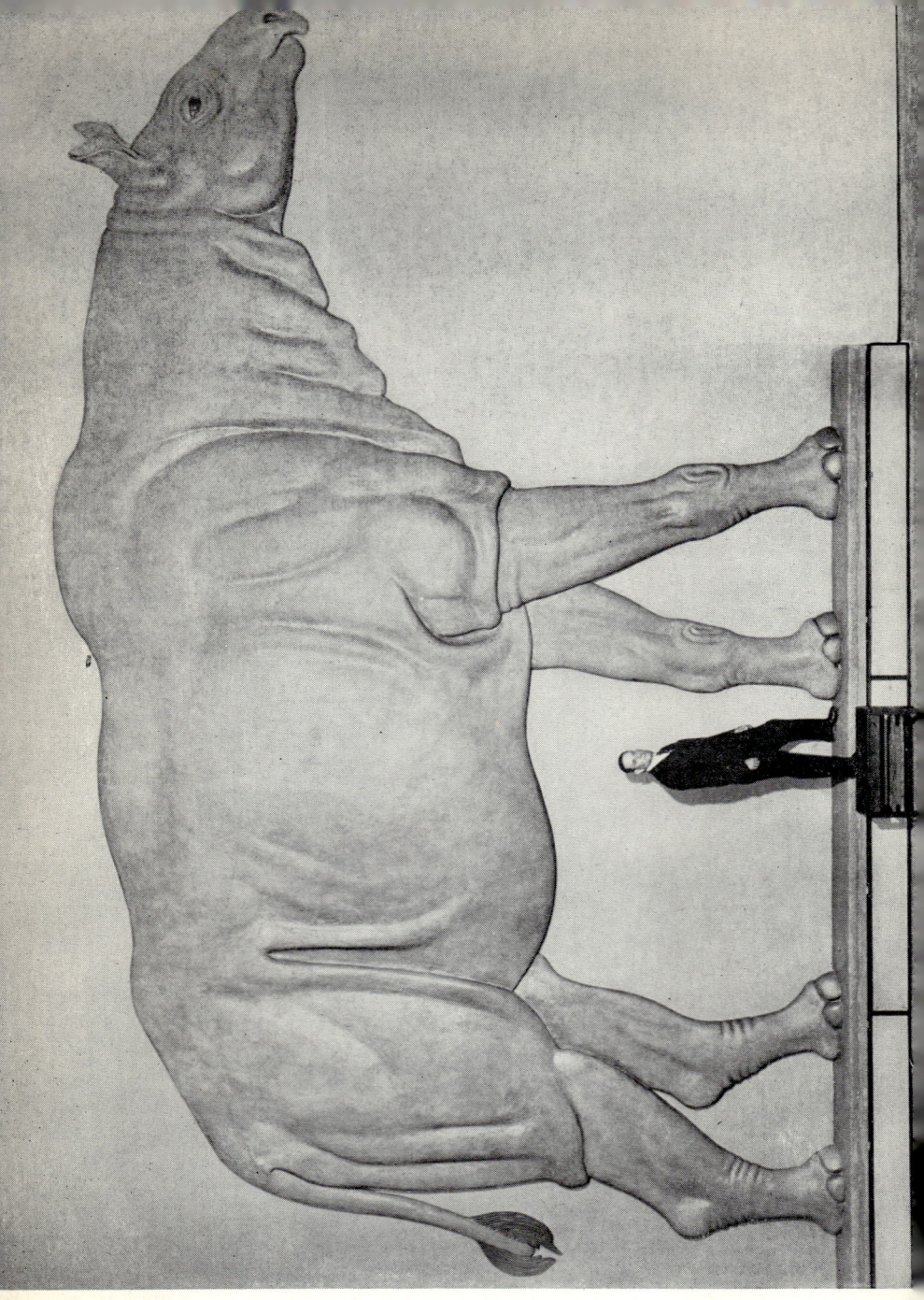

Above: Sculptor John Hope stands next to his bas-relief restoration of the giant fossil mammal, the Baluchitherium. The bones of this prehistoric rhinoceros-like animal were excavated from the sandstones of the Gobi Desert in Mongolia.

Paleontologists Barnum Brown, Roland Bird and Erich Schlaikjer examine the restored skull of a giant fossil crocodillian. The small skull is that of a full-grown modern crocodile.

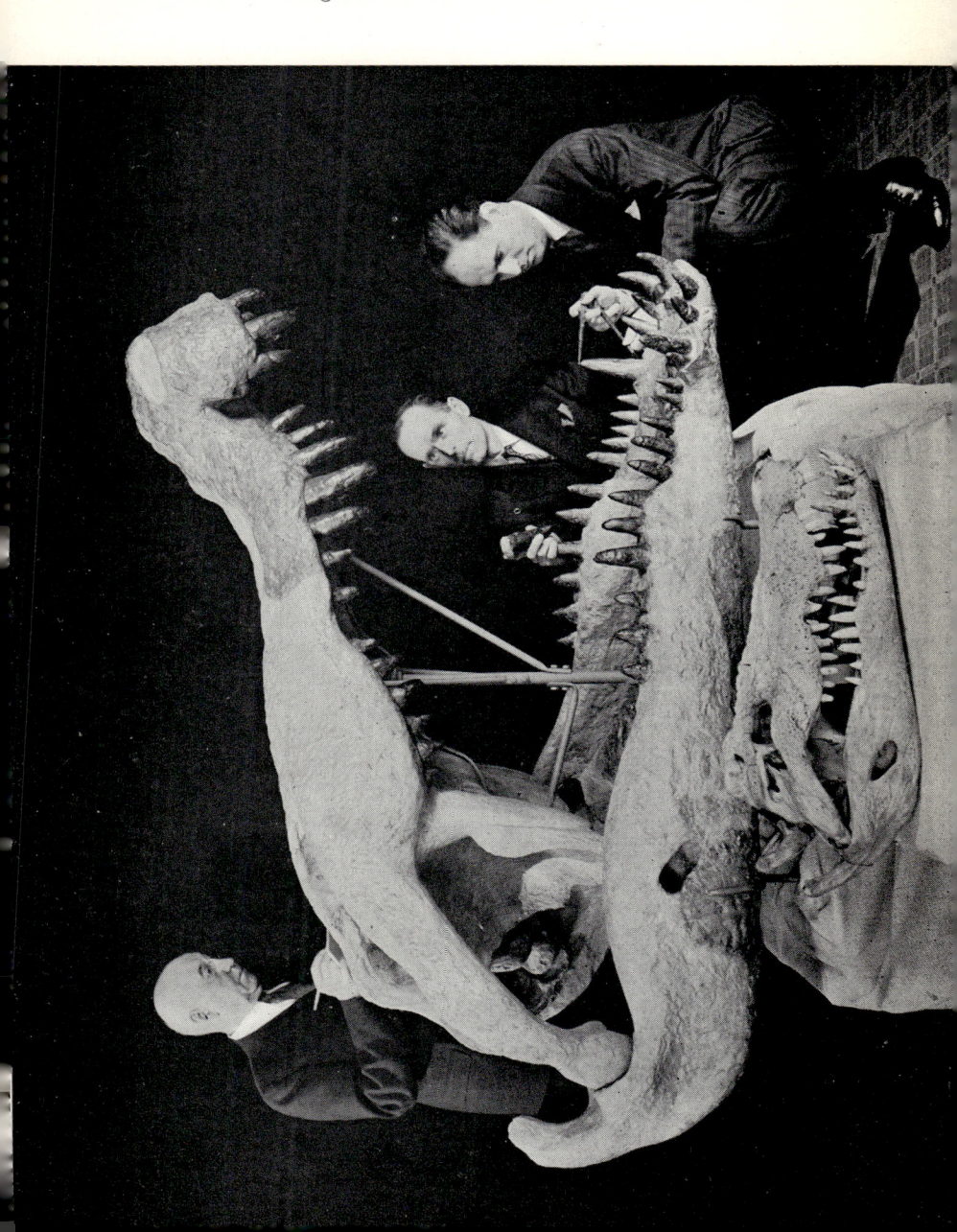

Smilodon, the prehistoric sabre-toothed tiger from the famous Rancho La Brea Tar Pits near Los Angeles. This great cat lived 20,000 years ago and was the most highly perfected killing mechanism in the history of mammalian evolution. A painting by Charles Knight.

slightly in appearance so far as the stars are concerned in thousands of years. One of the globes projects the stars of the northern hemisphere of the sky, the other those of the southern.

There are the equivalent of sixteen projectors in each globe, each one made to project the stars of a given section of the sky. The slides in these projectors are made of copper through which tiny holes have been punched, each one allowing a beam of light to be thrown upwards onto the dome to represent a single star. To cut these holes is a precision job of extreme delicacy as they must vary in size depending upon the magnitude of the star to be represented, and their placement must be carefully worked out so that when all are projected they will give the natural appearance of the star-filled night sky. Light is provided by a powerful central lamp mounted inside each globe. The copper slides are not square or rectangular but are so shaped that their images fit together on the dome like cells of a honeycomb.

In the frames which support the globes, or if you like, in the handle of the giant dumb-bell are seven special projectors. Five of these are for the five planets which are visible without the aid of a telescope. The other two project representations of the Sun and the Moon. All of these seven projectors move on eccentric gears which have been so carefully and accurately cut that they do not need oil, but in lieu of ordinary lubrication are carefully dusted by a hand bellows at frequent intervals. According to Dr. Bauersfeld, the construction of the moving parts for these special projectors which show the "wanderings" of the planets, including their peculiar and periodical retrograde motions, the phases of the Moon and other variations in motion, were the most difficult of the many problems of construction.

Most of the control of the machine is in the hands of the lecturer who stands in total darkness at a console or control desk with dozens of switches and rheostats at his fingertips. Additional controls are in a special projection room outside of the main room. Here are standard lantern slide projectors, motion picture projectors and a sound system, all of which are used when the occasion demands, to give special effectiveness to the illusion.

When skillfully operated the machine can produce the apparent and real motions of the heavens at several different speeds, all of which are many times faster than reality. Thus it is possible to speed ahead into the future or back into the past and crowd a long astronomical story into a very short time. In addition to these movements through time the machine also rotates on its axis to reproduce the effect of travelling over the Earth. The latitude of the observer seated in comfort in the Planetarium can be changed at will. In a matter of seconds the visitor can, in effect, travel from the Equator to the North Pole where he may view the stars as though he stood with Polaris, the pole star, directly over head, and then be whisked to southern latitudes and see the Magellanic Clouds and the famous Southern Cross.

This machine can show the heavens, at least all of the heavenly phenomena that can be seen by the naked eye, from any given spot on the surface of the Earth and at any given time of day on any day from the present to 13,000 years B.C. or 13,000 years A.D. This 26,000 year span is the time that it takes the Earth to complete one wobble.

For the Earth does wobble. In addition to revolving around the Sun, it spins on its axis and as it spins it wobbles like a top which is running down. If a real pole stuck out from the north pole the outer end of it would describe a circle that would take about 26,000 years to complete. This gives an apparent motion to the other heavenly bodies as seen from the Earth which is called precessional motion.

This motion accounts for the fact that Polaris, our pole star, was not always the star nearest the north pole of the sky. When the ancient Egyptians built the Pyramids and oriented them to the stars, the star that was nearest the north pole of the sky was not Polaris but Thuban. Some 12,000 years from now the bright star Vega will be the North Star and the Southern Cross will be visible from New York.

The projection planetarium has been acclaimed by leading astronomers as the greatest invention ever devised by man as a visual aid in teaching. So realistic is it in operation that it is a source of great enjoyment and delight to its viewers. Of it Dr. Elis Strömgren who directed the Royal Danish Observatory in Copenhagen wrote: "Never was a medium of demonstration produced as instructive as this, never one more fascinating in effect, and certainly never one which appeals to everybody as this does. It is a school, theatre and film all in one, a lecture hall under the vault of the heavens, and a drama in which the celestial bodies are the actors. No description, no photograph, no drawing can possibly reproduce the overwhelming impression made by a demonstration in a Zeiss Planetarium."

In its first year of operation the Hayden Planetarium played host to over 850,000 visitors and since then, millions of New Yorkers and out-of-towners have thrilled to its wonders. Many people come back again and again. The programs are changed monthly and include shows that exhibit the seasonal changes in the sky, such as "Stars of Spring," and "Stars of a Winter Night," and other spectacles of the heavens, such as "The End of the Earth," "Trip to the Moon," "Color in the Sky." One of the most popular shows repeated year after year on request and improving each time as new devices are invented to supplement the machine's standard facilities, is the Christmas show. With appropriate music, with such amazing effects as the familiar skyline of New York fading into the hills of Judea, with a snowfall and Santa Claus, the show takes the viewers back in time to the supposed date of Christ's birth. Several famous astronomers' explanations of the famous Star of Bethlehem are demonstrated. These include the unusual configuration or apparent coming together of three planets (Jupiter, Saturn and Mars); a nova or new star; and the bright planet Venus. Though hardly scientific, this phase of the Museum's astronomical program serves well one of the major functions of the institution, that of popular education.

The serious astronomical aspects of each Planetarium show are carefully and accurately worked out. The projector has been used to teach celestial navigation and was used during the last war to train many men in the United States Navy. Courses in star-identification and navigation are still being given and the projector is in constant use.

The first thing visitors see when they enter the Planetarium is Charles R. Knight's large mural painting based on astronomical legends and myths of the American Indians. It shows the Sun God pursuing the Moon Goddess across the sky. In the upper right hand corner are the Pleiades which the Blackfoot Indians spoke of as the Six Brothers, in the upper left the Big Dipper and the North Star. In the lower left corner the Ancient or Original Man sits on a mountain sending the little animals down below the water to bring up mud out of which he makes the world in accordance with an Indian creation legend.

Two features of the exhibition on the first floor of the Planetarium are the collection of meteorites and the Hall of the Sun, the latter containing, in the form of a large Copernican Planetarium, an animated, forty-foot model of our solar system, with the planets running on tracks around the Sun. The major moons of the planets are shown in addition to the Moon which is the Earth's satellite. Altogether there are about twenty-eight moons in the solar system and the Hall of the Sun model shows the two moons of Mars, the Earth's Moon, four of Jupiter's and the rings and four of Saturn's moons. The models representing the planets are set in their proper relation to the Sun, being all relatively in the same plane as they revolve about it, and their relative sizes are indicated except in the cases of the giant planets of Jupiter and Saturn. Mars is its familiar reddish color, Jupiter the yellow giant has its bands, and Saturn its famous rings.

The planets all are geared to revolve around the globe in the center which represents the Sun at speeds which represent their actual relative speeds. Thus little Mercury zips around the Sun about four times to every one revolution of the Earth. Mars takes almost twice as long to complete its orbit as does the Earth, its year being 687 Earth days in length. The circular walls of the Hall of the Sun have been decorated with pictures of the twelve signs of the Zodiac, the constellations which seem to be a background for the parade of planets.

On the floor of the Sun Hall is a beautiful reproduction in terrazzo of the famous Aztec Calendar Stone in Mexico City. The original stone symbolized the Sun, which to the Aztecs was the most important of the heavenly bodies. Surrounding the center figure of Tonatiuh, the Sun God, are symbols which commemorate the four great past epochs of the world and the present epoch. The Museum possesses a cast of the original stone which may be found in the Mexican Hall on the second floor.

The first floor corridor houses a collection of astronomical instruments, sun-dials, compasses, and astrolabes. This fine collection covers almost the entire range from ancient Chinese instruments through the fine metal instruments made in the middle centuries in France and Germany, down to the very accurate instruments which play such an important part in modern navigation.

Many fine astronomical paintings and transparencies adorn the walls of the Planetarium corridors and there is a set of dioramas which show famous observatories, old and new, including a model of the largest telescope of all, the famous Mt. Palomar 200 inch reflecting telescope.

An exhibit on the second floor, which is not astronomical yet one that intrigues many visitors is an exact duplicate of the famous time capsule that was lowered into the ground on the site of the World's Fair in Flushing Meadows, New York, a specially built non-corroding metal cylinder packed with items chosen to give a citizen of the earth 5,000 years from now a concept of what life was like in 1939. The duplicate cylinder in the exhibit is cut open to show its construction, and copies or duplicates of all the items which went into the original time capsule are displayed. The original rests some fifty feet below the surface in a specially built concrete shaft. A marker at the surface shows its location and books telling of its location have been printed and are stored in important libraries and other places where it is hoped they may be preserved.

On December 18, 1946 a special meteorological exhibit was opened to the public as a memorial to the late Lt. (j.g.) Joseph Prentiss Willetts who, with twelve members of his crew, was killed while on active duty in the Naval Air Service on August 18, 1943, probably because of unfavorable weather conditions. The exhibit was endowed by Lt. Willetts' family "to provide primary education in the field of weather, in the hope that this knowledge will save the lives of future airmen." The exhibit consists of a series of graphic models illustrating important weather phenomena and a "theme shaft" which depicts a cross section of the atmospheric envelope which encloses the Earth.

One of the graphic weather models by Eric Sloane depicts a tropical hurricane. Because the sea of atmosphere in which we live is invisible, most of us think of it as a void and we are no more conscious of it than a fish is of his own watery environment. Really we are more atmospheric-animals than we are earth-animals. We are inseparably bound to the medium of air. Since we live at the virtual bottom of this ocean of air we are adapted to it just as deep-sea animals are adapted to the ocean bottom. At the bottom of our "atmospheric ocean," that is, at sea level, the weight of the air exerts a pressure of fourteen pounds per square inch upon our bodies, furnishing us with the life sustaining oxygen without which we would perish, compressing one-half of the atmosphere into a space the upper limit of which is 18,000 feet above sea level, so that the air, its oxygen content constantly diminishing, as the distance from earth increases, becomes so rare at 18,000 feet above sea level that planes flying above that level must use pressurized cabins.

Air pressure also varies locally over the earth's surface because of the heat of the sun, the prime cause of all weather. The uneven heating of the air over various parts of the earth's surface sets up currents of air and produces unequal pressures. Some air is always rising, some falling, some advancing and some retreating. When extremes of air pressure are produced, violent weather results.

The model of the tropical hurricane in the Planetarium's weather exhibit shows such a disturbance. Pressure at the central core or "eye" of the hurricane is extremely low. Around this core circle winds that may reach velocities of over 100 miles an hour. The model shows how above the "eye" of the hurricane the sky is only partly overcast and wind velocity there drops to a minimum, producing the often misleading dead calm in the very center.

When the center of a hurricane moves directly over an inhabited spot, the inhabitants really feel the effects of two hurricanes. If the first fierce winds move from east to west, then after the calm center passes, fierce west to east winds are felt. This occurs because in the northern hemisphere hurricanes revolve counter-clockwise. In the southern hemisphere the order of wind directions is reversed because hurricanes south of the equator revolve in a clockwise manner.

Their paths are often erratic but generally speaking they move from south to north above the equator and from north to south below the equator.

The model also shows very graphically the size and the height of the hurricane. Usually the top of the disturbance is at an altitude of about 40,000 feet. Above this the weather is clear.

Another model, depicting the birth of a cumulus cloud, demonstrates one of our most familiar weather phenomena. Other models show the invasion of a cold front, a warm front, the world of clouds, and the formation of rain.

Since New York is not a good location for an astronomical observatory, no permanent facilities for examining the heavens with a telescope have been set up at the Planetarium. Usually, the sky over the city is quite smoke-filled and the many lights of the great metropolis make the night sky over the city too bright to see any but the most brilliant stars. During World War II when the "brown-out" kept city lights to a minimum, its inhabitants who looked skyward at night must have thought that hundreds of new stars had made their appearance.

Many New Yorkers, who have never seen a star-studded night sky in all its glory from some location out in the country away from the bright city lights have been puzzled by the multitude of stars they see in the Planetarium sky. On a clear night, in a rural location, most observers can see approximately three thousand stars within the limits of our horizon. All told there are about 9,000 stars which can be seen by the naked eye, but not from any one spot on the earth's surface. To see them all, a person would have to travel from New York to a point far enough south so that he could see the South Pole of the sky. But he can see representations of them all in a Planetarium.

Those who scan the heavens with the most powerful telescopes know that there are billions and billions of stars in space. When asked how many stars there are a great astronomer answered, "There are more stars than all the grains of sand on all the beaches of the world."

Astronomy can be a fascinating hobby. The American Museum of Natural History has long cooperated with two organizations composed of amateur astronomers, the Amateur Astronomers Association of New York, a group of adult enthusiasts, and the Junior Astronomy Club. For many years, under the auspices of the Amateur Astronomers Association, a class in telescope making has been conducted in a special room in the basement of the Planetarium. No giants of Palomar or Mount Wilson have been constructed here, but many very efficient and professional looking six-inchers, completely constructed by amateurs working under the guidance of a skilled instructor, have been carried out of the Planetarium basement by their proud creator-owners. The mirrors were patiently ground by hand and the loving care and patience which went into their manufacture rivalled that lavished on the construction of some of the world's largest professionally produced 'scopes. Both amateur societies hold frequent meetings and have had some of the world's best-known astronomers deliver lectures to their members.

Although not engaged on a research program, the staff of the Planetarium has frequently gone on expeditions to photograph and record eclipses. On one of these trips, the Junior Astronomy Club, then still in its early years, managed to appropriate enough money from its limited treasury to send along one of its more promising young members. Amateur societies have long been responsible for providing the stimulus that often starts some young man or woman on a professional career in science. The Junior Astronomy Club has done its share of proselyting for astronomy.

Since total solar eclipses do not occur frequently, and since they are seen as total eclipses only from a relatively limited area of the earth's surface, astronomers usually combine forces and sometimes make long journeys from home just to observe a phenomenon that lasts but a few minutes at best. Long preparation must be carried out so that everything will be in readiness for the very short observation period involved. Sometimes, for the ground parties, all of this preparation goes for naught when at the moment of the beginning of the eclipse, an errant cloud slips between the sun and the observers.

Artists often go on these expeditions to get the sketches from which they can later produce paintings in oil of the eclipse. Obviously a model that cannot "hold its pose" except for a matter of seconds, is not going to allow an artist time to complete a portrait in such a scant "sitting."

One of the eclipse expeditions of the Museum's Astronomy Department was sent to South America to study and photograph the solar eclipse of June 8, 1937. In addition to Dr. Fisher, then Director of the Hayden Planetarium, and members of his staff, the party included Mrs. Isobel M. Lewis of the United States Naval Observatory. The members of the party took their posts at various stations in Peru. Most were on the ground with special cameras and equipment, some were prepared to go aloft by plane to photograph above any cloud formations which might obscure the spectacle. Dr. Fisher and his group were located at Huanchaco, Peru since there they could see the sun at its highest point during the progress of the eclipse.

One of the results of the expedition is the painting, "Second Contact and Shadow" by D. Owen Stephens, a "portrait" of the Solar Eclipse of June 8th, 1937 as it was seen by the Hayden Planetarium Eclipse Expedition to Peru. The painting was presented to the Planetarium by the Junior Astronomers Club, and hangs with other exquisite and accurate astronomical paintings by others, including Howard Russell Butler, James Perry Wilson, and Leonard M. Davis.

The Hayden Planetarium building differs from all other American Planetariums in the construction of its dome. The concrete outer dome is hemispherical, its outside diameter 81 feet. It is incredibly thin, varying from three inches at the top to three and one-half at the base. It rests on a concrete ring about twelve inches thick which in turn rests on the steel frame of the building proper. It is a shell, entirely self-supporting, without the aid of trusses, hangers or any other external features, and carries the fifteen-ton weight of the inner or projection dome which is suspended from it.

The Hayden Planetarium dome was not cast in the usual manner.

Ordinarily, concrete casting is done by pouring concrete into a mold of wood or metal and removing the two sides of the mold after the concrete sets. In constructing the Planetarium dome but one singlefaced wood form was used and cement was blown into place over it with a cement gun invented by the famous explorer Carl Akeley during World War I when Uncle Sam experimented in building ship hulls of cement. The beautiful African Hall in the American Museum is a memorial to the versatile Akeley, who also invented the Akeley camera.

The inner dome, suspended from the outer dome by small T-shaped bar anchors, is exactly 75 feet in diameter. This dome, too, is hemispherical and its base has been cut into an irregular horizontal line which represents a silhouetted skyline of New York City as seen from a hill in Central Park not far from the Museum. In the soft light of the dome room visitors are apt to think that the skyline of buildings has been superimposed upon the main dome but close examination shows them that it is actually cut into the base of the steel plates which make up the inner dome. Since the view of the skyline is from Central Park one can see the Empire State Building, the Chrysler Building, and other prominent structures that tower upward in the sky-scraper district. Looking westward one can see the familiar outlines of the east facade of the American Museum of Natural History. The inner dome is made of stainless steel plates which have been painted. Objectionable echoes would bounce back from such a dome if it were not for the fact that these steel plates are perforated with more than twenty million tiny holes about $1/16$ of an inch in diameter spaced about $1/5$ of an inch apart. Any sound which passes through this sieve is absorbed by special acoustic padding behind the dome. The holes do not interfere with the star images.

When in the course of operating the machine stars dip down to the horizon they disappear and do not continue to be projected on the lower walls or floor. This is because their light is cut off at the horizon line by a small moving shutter that works just as do the eyelids of a "sleeping" doll.

Since the buildings in the skyline have been cut out stars that have not reached the true horizon do not shine on the buildings but their light passes beyond the line of the dome and strikes a glossy black tilted screen which deflects the tiny beams of light downward where the observers cannot see them.

The fact that the inner dome is perforated by tiny holes makes a special effect possible. When the lights are off inside the dome, and a light is put on between the outer and inner domes, one can see anything between the domes that the light illuminates. On the "Trip to the Moon" show, the control cabin of the rocket ship that is supposed to take the audience to the Moon is constructed between the two domes and, when lighted up in the manner described, gives a remarkably realistic effect.

Shortly after the Planetarium was opened to the public there was a show for school children from the New York City schools. On this day the demonstration chamber was filled with youngsters from the third and fourth grades. Their knowledge of things astronomical was practically nil and they had never been to the Museum or Planetarium before.

When the lights slowly dimmed and the night sky was revealed in all its glory, a gasp of real amazement went up from the young audience. As the show progressed and the illusion grew many of these youngsters forgot they were inside a building and thought they had been taken outside and that it was night. During the course of the show a technician who had some work to do and who had forgotten about the audience in the chamber, climbed aloft on the catwalk between the domes. When he turned on his light between the domes some of the children were shocked and frightened to see suddenly a glow of light and then the figure of a man way up in the starry sky.

One frightened youngster turned to his older companion and grasping his arm, whispered, "What's that? Jimmy, what's that up there in the sky?"

Jimmy, not much older and just as much taken back by the sudden and mysterious apparition, was heard to comfort his companion, "I don't know, Tommy, but I think it might be God."

This is not a far-fetched story when one considers the amazing effect a Planetarium show has upon those who witness it for the first time. Many adults have been carried away by the demonstration and they have remarked afterwards that they were sure they were out-doors looking at the real night sky, so complete is the illusion produced in the darkened auditorium with its hemispherical roof.

Astronomy is the great inspirational science. For thousands of years people have gazed in silent wonder at the night sky, have grovelled in fear when eclipses suddenly darked the sun, or when great showers of meteors filled the night sky with streaming flashes of light. As man's knowledge of the heavens increased, the very universe seemed to be expanding before his eyes. More powerful telescopes showed to him that beyond what he had thought were the limits of the universe were stars and still more stars. What a challenge to man has been the mystery of the heavens!

Poets, artists and musicians have sought to describe the wonders beyond this planet Earth. On the wall of the entrance foyer of the Planetarium is this statement by the man who gave the Planetarium instrument to the Museum and whose name the building now in memory bears.

THE WORLD OF NATURAL HISTORY

"I believe the Planetarium is not only a place of interest and instruction but that it should give a more lively and sincere appreciation of the magnitude of the universe and of the belief that there must be a very much greater power than man responsible for the wonderful things which are daily occurring in the universe."

Charles Hayden

CHAPTER TWO

Beneath the Grass Roots

BENEATH THE GRASS roots lies the great silent and immobile kingdom of the minerals. Extending below the earth to depths unknown are miles and miles of rocks made up of minerals which, in turn, are composed of various combinations of elements. These elements, of which over 100 are now known to man, are the veritable building blocks of the universe. We search for and detect the rare ones on earth and by means of spectroscopic analysis, we discover the presence of similar elements in the Sun itself. The examination and analysis of those visitors from space, meteorites, has not disclosed any elements which are unknown on the earth.

The science of Geology and its branch Mineralogy have long been a concern of the American Museum of Natural History. The very first of the scientific departments of the Museum to be established and to be endowed with a full

Curator at its head was the Department of Geology. When the original staff of the Museum was made public it included the name of Dr. R. P. Whitfield as the Curator of Geology. Dr. Whitfield's branch of geological specialization was invertebrate paleontology (study of fossils of backboneless creatures), but his responsibilities were those of the entire field of geology for the Museum.

Before Dr. Whitfield came to the Museum in 1877, the Museum's mineral collections began with the purchase of the Bailey Collection of approximately 7,000 cabinet specimens which was acquired in 1874 for the sum of $4,000. It was stored in the Arsenal building and was transferred to the Geology Hall in the permanent building in 1882. The Museum had already owned a few mineral specimens which had been given to it by its friends but these were of little importance. The Bailey Collection was a fine start. Visitors to the Museum were really able to study minerals and to obtain considerable knowledge of their relations and relative importance as the Bailey Collection, though it was neither large nor noted for the brilliancy and rarity of its contents, was a very representative and useful one. With it also came some meteorites, and thus began the Museum's collection of "Shooting Stars."

In 1876, the Museum received the minerals which had been exhibited at the Philadelphia Centennial Exposition by the Governments of Canada, Spain, Brazil, Tasmania and New Zealand.

And in 1891 the Museum purchased the remainder of the great Norman L. Spang Collection, the bulk of which had been previously acquired by Mr. Clarence S. Bement, who was building up one of the finest collections in the world. This addition really gave the Museum's geological specimens a distinct rank among notable mineral collections. Some of the large and impressive specimens in the wall cases today are from the Spang Collection.

The real benefactor of the Museum's mineral and gem collection was the late J. Pierpont Morgan, one of the Museum's founders. In 1890 Mr. Morgan purchased for the Museum the first part of the gem collection which had been made up by the Tiffany Company for display at the Paris Exposition of 1889. There were some beautiful sapphires, topazes, beryls, garnets and tourmalines in this collection. To these, in 1900, Mr. Morgan added the 1,453 gems in the Tiffany Exhibit at the Paris Exposition of that year, more than doubling the size of the Museum's gem collection. Additional gifts from Mr. Morgan, Mrs. Edith Haggin De Long and other friends have made the Museum's gem collection first in size, value and importance in the country.

However, gems are but a part of the story of mineralogy. So far as making the Museum's entire mineral collection of great value, it was not the gift of gems but of the Bement Collection of Minerals which Mr. Morgan gave to the Museum in 1900 that made the Museum the owner of the leading collection of minerals in the country, if not the world. Clarence Bement was one of the most persistent of private collectors. It was his aim to build a collection which would ultimately contain one or more specimens of every variety known to man. Offhand, this might not seem like too difficult a task since there are less than 1,500 varieties of minerals. The difficulty is that the bulk of the earth's crust is made up of rocks which are in turn chiefly composed of combinations of about thirty kinds of very common minerals. All the others are relatively, and some exceedingly, rare.

It is a tribute to Bement's patience that his collection contained the best possible specimens available at the time of about 1,000 different kinds of minerals. It has been said that Mr. Bement would purchase a whole collection of minerals in order to secure a single specimen of unique value. He spared no expense and had some of the best mineral collectors in the country searching the world over for treasures for his collection. In addition to its priceless specimens the Bement collection also contained a series of meteorites representing almost 500 falls and finds.

It was the addition of this collection which Mr. Morgan presented to the Museum which caused the establishment of the Department of Mineralogy as a separate department with Mr. L. P. Gratacap as the first Curator of Mineralogy. Mr. Gratacap sought to continue along the lines which Mr. Bement had followed. But, except for gifts which were made in the form of specimens there was no opportunity for the department to add to its collection. What was needed was a fund with which to purchase specimens to round out the main collection. Money for this purpose was generously provided by friends of the department. Miss Matilda Bruce and Bonnie Wallace LeClear established endowments for the purchase of new specimens and in 1931 William Boyce Thompson willed his magnificent collection of showy mineral specimens to the Museum with an additional bequest of $20,000 for its installation and $50,000 for its maintenance.

In 1922 George Fisher Baker honored the memory of his friend and associate, J. Pierpont Morgan by providing for the complete remodeling of the mineral hall. Since that time it has been designated as the Morgan Memorial Hall of Minerals and Gems.

The general collection of minerals which Morgan Hall contains in addition to the Gem Collection is one of the finest in the world, ranking with that of the British Museum and the Jardin des Plantes. It occupies most of the space in the large hall; the larger and more imposing specimens are to be found in wall cases and the smaller specimens in floor cases nearby. The collection is arranged according to the classification of minerals and begins with a series of cases containing minerals which are elements, such as Gold, Sulphur, and Copper. The series continues with Sulphides, Haloids, Oxides, Carbonates, Silicates, Phosphates and finally Sulphates. A long row of double-sided cases running the full length of the hall contain the bulk of the Gem Collection. In other parts of the hall are special exhibits devoted to crystallography, economically important minerals, color in minerals, and the molecular structure of mineral substances. From time to time temporary exhibitions which feature some interesting and significant aspect of mineralogy are installed in cases near the center of the hall. There is also an outstanding collection furnished by the New York Mineralogical Club which contains representatives of approximately a hundred varieties of minerals collected within the limits of the City of New York.

It is the Gem Collection which attracts many visitors from afar. One of the prizes of the entire collection is the famous 563-carat "Star of India," the largest cut star sapphire in the world. Near it is the beautiful De Long star ruby, one of the finest gems of this kind in the world. It was given to the

Museum by Mrs. Edith Haggin De Long. It weighs 100 carats and was discovered in a Burmese ruby mine. Colored a peculiar orchid-red it reflects a six-pointed star as does the famous Star of India.

The crystals from which star gems of this type have been cut contain a myriad of minute hollow tubes. These tubes are distributed throughout the crystals with great regularity, parallel to the six sides of the natural crystal. When cut *en cabochon,* or dome-shaped, so that the rounded dome arches over the hexagonal pattern of tubes, these stones reflect the light from the interior as a six-rayed star.

The gems in the Museum's collection are arranged in a series which begins with the diamonds, the hardest of the precious gems, and continues with gems of decreasing hardness and decreasing value. In addition to the precious gems are many fine examples of semi-precious gems and ornamental minerals. Wherever possible the crystals are shown as they are found in nature and cut examples are displayed nearby. The diamonds in the bluestone matrix as found in South Africa are shown side by side with diamonds in the rough from South America and elsewhere. There is a tray of small diamonds especially selected to demonstrate the range of colors which diamonds display, surprising many who do not realize that diamonds occur in any color other than the familiar blue-white of the average stone.

Although minerals may appear as veins in some other mineral or rock, or in massive chunks which do not show any resemblance to crystals, all have crystal forms which are sometimes so perfect that it is hard to realize that they "grew" that way and were not shaped by some clever lapidary. But not all crystals of the precious stones provide gem material.

Even when one is of gem quality, it must be studied for a long time by a skilled gem-cutter before he can decide what size stones he can best cut from it. The more flaws in a crystal, the less valuable it is as a potential source of fine cut stones.

The Museum collection includes some very fine large natural crystals of ruby and sapphire, the corundum gems. They are not quite so hard as diamonds, having a relative hardness of 9 on a scale which diamonds top with a rating of 10.

Well-cut emeralds of good green color have commanded prices rivalling that of the best grade of diamonds. Emerald is but a variety of the mineral beryl, which owes its beautiful green color to what is actually an impurity—a small amount of chromium. The stone has a very low refractive index, which accounts for its typical lack of fire and brilliance, and usually has imperfections such as tiny cracks and inclusions. Flawless emeralds are thus very rare. When one is found, a perfectly transparent gem, it commands a high price, sometimes three or four times that of a diamond of equal weight.

The engraved emerald, given to the museum by Miss Cockcroft-Shettler, and one of the finest pieces in the Morgan Hall's display, weighs about 87 carats and originally came from Columbia, the principal source of emeralds and specifically from the Muso Mines, source of emeralds for the aborigines in the days of the Spanish Conquistadores. It was probably cut in Delhi, India, and used as a head ornament by some Hindu prince many years ago.

The Museum has many fine topazes. Although topaz gem crystals are found in a variety of yellow, orange, and blue shades, the orange-yellow shade resembling sherry wine is the most typical of the gem stone. Common usage has given the name to stones which are not topaz at all but simply have the characteristic yellow color. "Oriental topaz," for instance, is but yellow sapphire and "Spanish topaz" but an amber-colored quartz.

Topaz is almost as hard as emerald and specimens of high quality rank with the best of the semi-precious stones, the title of precious stones being reserved for diamonds, rubies, sapphires and emeralds. Most of the best colored topaz comes from Brazil although the stone is obtained from many other localities. The Museum possesses the largest gem quality topaz crystal ever found, a giant which weighs more than one-quarter of a ton. Of the cut specimens in the collection the show piece is an egg-shaped brilliant cut blue topaz from Japan, with 440 facets and weighing 1,463 carats.

Occupying a place of honor in a hall filled with excellent examples of the best in lapidary art is a chalcedony figurine about ten inches high, the gift of Charles Lanier. This carving, representing the high skill of the lapidary, Georges Tonnelier, is perfect down to the tiniest detail. Thousands of other examples of semi-precious minerals, some cut in exquisite taste and all in all a world tour of gem craft from prehistoric times to the present, are in the Morgan Hall and are literally the treasures of a house of treasures.

Almost hidden away in the West Tower Room on the fourth floor of the Museum is one of the finest collections of carved jade, ivory and amber to be seen in any museum. Assembled by its donor, the late Doctor I. Wyman Drummond, it is really a group of collections, each one of importance and charm. The carved jades alone form a rich and well-balanced series, containing examples from all periods and covering a cultural range of over thirty centuries of oriental history. In the center of the room is a very unique composite piece of the purest white jade, consisting of an assemblage of elaborate carvings which have been fitted together. Thirteen pieces of jade fit together to form a round group some 12 inches in diameter. Twelve pieces have been cut to fit around a circular center piece. Each of the twelve outside pieces is carved with a representation of one of the twelve terrestrial branches corresponding to the signs of the zodiac. Looking not unlike a beautifully decorated birthday cake this masterpiece was actually a gift to the Emperor Kien Lung from the officials of his court on the occasion of his 50th birthday in 1768.

Another fine piece is a slender Joo-I scepter in white jade also of the Kien Lung period. This "Scepter of Good Luck" has carved on its long handle the figures of the Eight Immortals of Taoism, each one with some characteristic object in his hands such as the flute of Han Hsiang-tzu, whose marvelous tones could make flowers grow and blossom in an instant. No one is quite sure as to the origin of the shape of the Joo-I scepter or has its original use been satisfactorily explained. Some authorities think that the shape symbolizes the Fungus of Immortality, others viewing the scepter sideways, think it represents the Chinese constellation of the dragon, and still others think a survival of a primitive kind of divining rod. Back in the thirteenth century a Chinese archaeologist is reputed to have said of the Joo-I scepter that it was used to "point the way and to guard against the unexpected," that is, that it was used for gesticulation and for self-defense as a sort of blunt sword. In more recent Chinese custom the scepter's use is purely ceremonial. It is presented as an auspicious token of friendship, a symbol of good luck. Its very name indicates its versatility for it means "in accordance with your wish."

Among the valuable and interesting items in the Drummond Collection are hundreds of little carved ivory buttons or bobs called netsuke and a series of exquisitely carved snuff-bottles in jade, glass and semi-precious minerals.

The Morgan Hall also contains many oriental pieces of jade carving including a very unusual necklace and pendant all of which has been carved out of one piece of jade without a break.

The American Museum of Natural History thus presents to the visitors to its Mineral and Gem Collection a fine series of specimens running the gamut of the whole classification of minerals and a lavish display of priceless gems and outstanding examples of the art of the lapidary.

Ever since the Earth was created by the tremendous celestial cataclysm which removed it from its solar parent it has been steadily undergoing a change. The long process of cooling which has resulted so far in giving it a rocky crust still continues within it. The gases which exuded from the seething planet to form its atmosphere and the water which condensed and streamed down upon the rugged surface time and time again helped to mold its shape. The great stresses and strains that were the result of the solidification of the new planet have continued until the present, and will continue for many centuries.

The story of the changing earth, of the formation of its outer layer of rocks, of the constant effect of the wear and tear of its surface by the elements and by the vast movements within the planet itself, is the story of geology which comes under the branch of that science known as *physical geology*. In its structural aspects this story deals with the form, arrangement and internal structure of rocks. It deals with minerals and their arrangement and association in rocks and ores. It deals with the occurrence, distribution and history of the three major kinds of rocks, the igneous, the sedimentary and the metamorphic. In its dynamic aspect, physical geology deals with all the processes and agencies which work within and without the earth's surface and which alter its appearance. Dynamic geology is concerned with the study of such phenomena as rock-weathering, underground waters, glaciation, diastrophism, vulcanism, metamorphism and gradation. It considers the work of wind, water, snow and ice, plants and animals, and man himself as all these agents do their part in sculpturing the planet's surface. It deals with such monstrous forces as are expended in earthquakes and volcanoes. It concerns itself with the modifying aspects of the glaciers now and in the past.

There is also an historical side to physical geology. These changes have been going on for the several billions of years of our planet's history. The geologic historian attempts to produce a chronological account of the earth's history of geologic change.

In general, the Museum's exhibits designed to contribute to a study of our changing planet consist of two kinds of displays. On the one hand, there are the actual bits of the earth's crust, the rocks and ores, the volcanic products, the glacially-rounded and scratched stones that are tangible evidence of the process of earth formation and modification. These alone would prove dull except to the interested student. The other kind of display consists of models, charts, and reproductions of outstanding examples of some of the most evident and striking results of geologic forces.

In the very early days of the Museum's career, the types of exhibits displayed were limited to those of the kind first mentioned, namely collections of rocks and minerals. In 1869 the Museum was given its first collection of rocks, an important collection which illustrated the geology of Palestine and Sinai. On occasion, some of the duplicate specimens in the Hall Collection of fossils were successfully traded for specimens which the Museum did not have, as when, in 1878, in exchange for two series of Hall duplicates, the Museum obtained from Professor C. H. Hitchcock a complete series of Vermont and New Hampshire rocks. This collection of about 2,000 specimens illustrated the geological survey of those states which had been carried out under Hitchcock's direction. In the same year the Museum became the depository for the collection of rocks and fossils which had been gathered by the United States Survey of the 40th parallel. In addition to fossils, the collection contained 3,880 rocks and some 2,800 microscopic sections of rocks.

In 1886, Morris K. Jesup, then President of the Museum, presented his institution with a complete duplicate of the series of building stones which had been collected under the direction of the 10th Census Commission. This series consisted of 1,053 four-inch blocks, polished on one face and variously dressed on the other sides. The specimens represented nearly every State and contained samples of all the different kinds of rocks used for ornamental and building purposes. Since that time the Museum has added to its collection of rocks and economically important ores.

BENEATH THE GRASS ROOTS

One of the chief features of the Museum's physical geology displays is a series of fifteen pictorial relief models each representing an area within the United States which exhibits striking results of certain geologic forces. These include models of the Bright Angel section of the Grand Canyon of the Colorado River, the Niagara Falls region, Crater Lake in Oregon, Pike's Peak, Yellowstone Canyon, Yosemite Valley, Mt. Washington in New Hampshire and seven other localities. The models are all worked out carefully as to scale and the front of each represents a cross-section of the rock formations of the area involved. The groups or models, each against a painted background, are enclosed in cases similar to those which contain the animal habitat groups.

The Grand Canyon of the Colorado, discovered in the early sixteenth century by the soldier-explorer Cardenas, is one of the most amazing examples of the erosive power of wind and water. It is more than 300 miles long, thirteen miles wide at its widest and a mile in depth from the brink of the canyon to the river bed below. The great cut reveals a veritable book of earth history, the pages being represented by the successive layers of rock solidified during different ages in the formation of the earth's crust. At the very bottom of the canyon, now the bed of the Colorado River are exposed the oldest of the formations, the hard granites and gneisses formed countless ages ago. Later the region sank below the sea and, one after the other, layers of sandstone, limestone and shales were deposited on top of the older granites and gneisses. These sedimentary deposits built up to a thickness of more than 4,000 feet. After more millions of years had passed, these same rocks were raised above sea level and were exposed to the erosive effects of water, wind and frost. Once again the region was submerged beneath the sea and more marine deposits were laid upon the older formations. Changes such as this continued through the ages and about a million years ago the whole area was again above sea-level and the excavation of the present canyon was begun by nature. This cutting of the canyon is still going on but it is quite slow because the river now runs on the harder and older igneous granites and gneisses at the bottom of the canyon. Glancing upward from the bottom of the canyon one can see the layers of sedimentary rocks which unfold many thousands of years of the story of the earth's past.

The Museum's model, one of the finest in the physical geology exhibit, represents a part of the canyon about sixteen miles wide from west to east and thirteen miles from north to south.

The industrial significance of geology is represented in Geology Hall on the fourth floor by a model, eight by twelve feet in size, which shows in scale the surface features and buildings over four of the main mines belonging to the Copper Queen Consolidated Mining Company of Bisbee, Arizona. The side of the model reveals 1,200 feet of a vertical section of the mine, shows the way in which the ore is removed and how new deposits are discovered. In cases close at hand are specimens of copper ore, minerals and rocks from the mine and surrounding area. In a case by itself is a great block of malachite and azurite taken from the original "Queen" mine. This beautiful blue and green specimen weighs 9,000 pounds, and as the ore is quite rich in copper, contains almost two tons of that useful metal.

The Geology Hall also contains models of two caves. The

larger, which visitors may enter, is a representation of a chamber in Weyer's Cave, Virginia, a limestone cave with large stalactites hanging from its ceiling and stalagmites protruding up from its floor. The smaller is a reproduction of a very beautiful cave discovered at the Copper Queen Mine. It was formed by the dissolving action of water traveling along joints in the limestone. Its walls, roof and floor were subsequently covered and coated with incrustrations, stalactites and stalagmites of calcite, some of which are a dazzling white while others are tinted green with copper salts or a delicate pink with manganese compounds.

Additional material of economic importance is to be found in the Museum's Hall of Economic Geology and Petrology on the third floor. Here are more mine models and an exhibit of coal and its many by-products. Here also is the bulk of the Museum's collection of rocks, ores and building stones and its exhibit on volcanoes and earthquakes.

The Museum's instrument for recording earthquakes, the Mainka seismograph, is on the first floor, its main frame resting upon a concrete column which in turn rests upon the bed rock. Earth tremors travel through the concrete column to the machine and are recorded as thin-lined graphs on smoked paper. Under normal conditions, the lines, drawn by a special stylus on blackened paper mounted on slowly revolving drums, are straight. When a severe tremor occurs the delicate inscribing needles move from side to side and produce a zig-zag line. The intensity of the quake is indicated by the extent to which the line veers from a straight path, the time of occurrence and duration by a little stylus which makes a mark on the paper for every time interval.

The exhibit devoted to earthquakes and volcanoes includes paintings of the more famous of these catastrophes and mute reminders of their force and destructive powers in the form of burnt and blackened misshapen bits of debris found in the remains.

For example, there is a twisted, partially melted, street sign on which the observer can barely make out the name "Rue de Victor Hugo." The story behind it is that of one of the most destructive volcanic eruptions the world has ever known. It occurred in May, 1902 on the island of Martinique in the West Indies. The island's largest city was St. Pierre, a town of 40,000 inhabitants nestled snuggly at the base of Mt. Pelée. Affectionately called "Papa Pelée" by the townspeople, it had been peacefully dormant for so long that no one feared its known volcanic nature, but considered it only as a picturesque backdrop for the town, and an old friend which offered its hospitable green-clad slopes for holiday outings.

But late in April of 1902 Pelée began to rumble uneasily in its hidden depths, and small wisps of smoke floated up from its long quiet crater. No one paid much attention to it at first. On May 4th when a smoking stream of mud and lava burst through the top of the crater and plunged into a nearby river valley killing twenty-four workmen in a sugar plantation everyone thought that there was no further danger. When it erupted violently on May 8th, the town's inhabitants were caught by surprise. A great cloud of poisonous and searing hot gas rolled down the slopes of Pelée and snuffed out the life of every living thing in St. Pierre save one lone occupant of the town's jail, a condemned murderer who while awaiting his execution had become ill with a fever and had found relief from the burning fever by sleeping in the cool cellar. He was there when Pelée erupted and was thus spared. Shortly after this disaster which claimed 40,000 lives, the Museum sent Dr. Edmund Otis Hovey to the site of destruction to study the ruins and to learn what he could of the volcano.

This interest in volcanoes is still continued by Dr. Frederick Pough of the Museum's Geology Department. Dr. Pough has made many trips to the site of Parícutin, in Mexico, the earth's youngest volcano. He has watched Parícutin almost from the moment of its birth to the present day, making a pictorial record of its growth and activities and studying its products.

Parícutin has been of tremendous interest to all volcanologists as it is the only volcano that has been born in recent times, and is thus the first in history to come under scientific investigation almost from the moment of its birth. It lies in a region of recent volcanic activity in the western Sierra Madre mountains about 200 air miles west of Mexico City. The first indication of its coming was a series of earthquake shocks in the region, about two weeks before its birth. The only witness of its natal struggles was a farmer who, while plowing his cornfield on the afternoon of February 20th, 1943, saw a spiral of steam curl upward from a small depression in his field. During the night tremendous explosions rocked the rural community and when the sun rose the next morning it shone through a heavy mist of steam and smoke and dust, and in the cornfield was a volcanic cone already more than fifty feet high.

The activity that followed during the next six months was awe-inspiring. Puffing and exploding every few seconds the new volcano ejected cinders, ash and bombs of solid rock which some observers estimated came out at the rate of 2,700 tons a minute. Thus, the cone built up rapidly. At the end of only six months it had reached a height of 1,000 feet. After four months lava appeared in the crater, burst out of the sides of the cone, and by the end of the volcano's first year had engulfed two villages. The peak was now 1,500 feet high and the ash and gas which had issued from its vent had destroyed all vegetation for twenty miles around. More than 8,000 people were forced to leave their homes, and a large area of land was laid waste. The activity has now calmed down almost to nothing compared to the earlier commotion and no one knows how long it may continue to be active. The nearby volcano El Jorullo, now extinct, continued to be active for 19 years after its birth in 1759.

The Museum exhibits on the volcano Parícutin and the volcano Pelée is flanked by exhibits on the classic eruption of Vesuvius which buried Pompeii, and that of Krakatoa.

Volcanoes and earthquakes are often associated. Many times the volcanic eruptions are themselves the cause of earthquakes. In other cases, like that of Parícutin, the earthquake shocks come before the eruption. Many earthquakes are of deep-seated origin, due to shifting and settling of great blocks of the earth's crust. The area of volcanic activity today is almost identical with the area of the most frequently occurring earthquakes. Generally speaking, it surrounds the Pacific Ocean and its activities are felt along the Pacific coastlines and on the islands dotting this great ocean.

The greatest single disturbance the modern earth has ever felt—greater even than would be a combination of all the atom bomb discharges to date—was the Krakatoa explosion in 1883 which blew away the entire central island (18 square miles) of the Krakatoa group between Sumatra and Java. The shock was felt right through the earth's diameter. Bits of pumice and dust were blown twenty miles into the air and the currents of wind in the upper air carried some of the dust completely around the earth. For a long time after Krakatoa, people as far away as the United States saw remarkably colorful sunsets which were caused by the dust from Krakatoa that remained in the atmosphere.

The Geology Hall on the fourth floor also contains the Museum's exhibit devoted to Invertebrate Paleontology under the supervision of Dr. Norman D. Newell and Dr. Otto H. Haas. Here are the fossil remains of the backboneless creatures that inhabited the earth in its early days. As mentioned earlier, the famous James Hall collection was the keystone of the Museum's invertebrate fossil collections and has been added to through the years. In 1873 the Museum acquired the Holmes Collection of South Carolina fossils and the United States Survey of the 40th parallel collection contained about 3,500 fossils.

In addition to invertebrate fossils the Geology Hall also contains the Museum's paleo-botanical specimens. Outstanding among these is a section of a prehistoric conifer tree over forty feet long, and several large stumps of fossil trees found in an anthracite coal mine under Scranton, Pennsylvania. These collections are arranged both in order of the age of the specimens and in terms of their zoological and botanical classifications. There are many fine examples of trilobites which flourished in the Devonian seas hundreds of millions of years ago. Along one side of the hall is a series of eight models which show successive stages in the geographic development of the North American continent from a prehistoric time when much of it was under water, to the present. The fossils in this hall present a parade of invertebrate life ranging from that of the ancient Cambrian seas down to that of recent times. Many types ancestral to our present forms can be seen preserved by nature through millions of years.

The search still goes on and bit by bit man is learning to put together a picture of life long ago. Some of the parts of this picture will never be found, as fossils are, in a way, accidents. Most animals and plants disintegrate after they die and are soon lost for all time. Only a comparatively few living creatures of the past, and too often only parts of these, are preserved in what we call fossils through the agencies of water and sand, or record their passage by tracks or imprints on some ancient mud which becomes fossilized and so is still discernible. The American Museum's men of science have expended a tremendous amount of time hunting for and studying fossils in their effort to probe the past.

CHAPTER THREE

Probing the Past

ONE DAY a rancher from Alberta, Canada happened to be visiting the paleontological laboratory of the American Museum of Natural History. He looked over some old bones that were resting on one of the tables and remarked to Dr. Barnum Brown that "bones like that are six for a nickel out my way." This chance remark led to one of the lucky strikes of the Museum's fossil hunters for it took Dr. Brown to Alberta where he discovered a locality rich with the fossilized bones of dinosaurs.

Not all of the Museum's fossil collection, the largest in the world, was obtained so easily or so quickly. Paleontologists, men who study life in the past, have had to work long and hard, covering many thousands of miles of rugged territory, to bring their prehistoric quarries back to the Museum in New York. Once back at the Museum, long months, often years, are spent by the preparators in the paleontology laboratory wresting these bones from the rock matrix which contains them and then the paleontologists themselves must spend long hours studying and measuring, comparing and describing, these old creatures that science is revealing to public knowledge after so many millions of years.

The Museum has been probing the past almost since the beginning of its institutional life. The first curator was an invertebrate paleontologist and the very first significant collection the Museum obtained was the Hall Collection of fossils. Despite the fact that a great and important part of the Museum's vertebrate fossil collection came as a result of the purchase of the Cope Collections and others, much of the Museum's fine accomplishment in this branch of geology can be traced to the men who served on the staff of its paleontology department and who spent years in the field collecting its specimens. Today, the bulk of the additions to the collection are the result of field parties the Museum sends out.

It was the intent of the Department of Vertebrate Paleontology when it was established in May, 1891 by President Morris K. Jesup and headed by such men as Henry Fairfield Osborn and William Diller Matthew to concentrate at first on fossil mammals. Its plan was to "form representative series from the successive horizons of the West" in order to present a "historical development of the Evolution of the Mammals in North America." This plan was broadened subsequently and the department's interest was extended to all of the fossil vertebrates.

The Museum's paleontological interests have been world wide for many years and it has maintained a relationship with Columbia University which has resulted in the training of many of the leading authorities in this country and abroad. The ultimate results of the Museum's scientific work are usually to be found in the published studies of the men who form its scientific departments. The scope of the work may be judged by the fact that when the fossils in the Museum's collections were scientifically described, the descriptions filled over 15 volumes of 1,000 pages each in the American Museum of Natural History Bulletins and Novitates. More elaborate studies on fossils in the Museum's collections fill many of the American Museum Memoirs and Special Publications. In addition to the Museum publications, numerous papers on these collections have been printed elsewhere.

Fossils in the Museum's vast collections fill six exhibition halls and the great study collections of the paleontology department seem to be everywhere in the non-exhibition areas. They overflow into the corridors of the fifth floor, they are stored in the attic and in the wide basement corridors and some of the unpacked material has had to be kept under tarpaulins in the service yards. They come from every continent, from the far north above the Arctic Circle to the southern-most outpost of Antarctica. Most of the specimens however hail from North America, particularly the rich fossil beds of the West.

After the Age of Invertebrates had given way to the Age of Fishes, some 500 million years ago, the prehistoric world was to see the advent of some of the largest and most bizarre of its creatures. In the earlier periods of this age the first of the vertebrates were rather small and related to the sharks and Ganoid fishes similar to the sturgeon of today.

The Bashford Dean Memorial Exhibit of Fossil Fishes on

the fourth floor with the rest of the Museum's exhibits of fossil vertebrates, is a memorial to Dr. Bashford Dean, the Museum's first Curator of Icthyology when that department was established in 1909. This versatile scientist was an authority in another totally unrelated field, that of armor, and was for many years the Curator of Armor at the Metropolitan Museum of Art. At the Museum his special field of interest was the armored fishes of past ages.

The entrance to the Fossil Fish Hall is marked by the dramatic exhibit of the reconstructed jaws of the giant fossil shark, *Carcharodon megalodon*. The great open jaws, nine feet across and studded with the actual fossil teeth set in place, seem wide enough to crush a jeep. The teeth, about 4¼" in height, are approximately four times the size of those of the white shark, the largest of the man-eating sharks extant today. The giant fossil shark whose teeth were found in deposits of marine fossils in the southern seaboard states of Georgia and South Carolina must have been the most voracious and destructive creature of the great seas that covered the Eastern part of the United States at the time. It is estimated that these monsters must have been almost fifty feet long.

The oldest of the specimens of fossil fish in this hall are the primitive Ostracoderms from the Silurian and Devonian formations. These jawless creatures lived more than four hundred million years ago and our modern bony fish, the Teleosts are a far cry from these primitive forbears of theirs. Ostracoderm means "shell-skins" because these early chordates had their heads and chests covered with a shell or shield of a hard substance related to dentine.

One of the great armored fishes that Dr. Dean loved to

study is restored and has been hung from the ceiling in the act of swooping down upon some of its smaller contemporaries. This model represents *Dinichthys* from the Devonian of Ohio. In a nearby case is the original skull of the fish, one of the prizes of the collection. *Dinichthys* was one of the many species of the *Arthrodira* or "joint-necks," which had a pair of joints connecting their bony head pieces with their shoulder plates. In their day they must have been tyrants. The nearly naked fin-fold sharks upon which *Dinichthys* might have fed demonstrate a very interesting stage of the evolution of fins. In these early sharks the skeletal rods which support the primordial fin-fold had begun to push outward well beyond the line of the body. In the long process of evolution we find that external limbs represent a later development and these fins of the fin-fold sharks represent one of the early stages in the development of body appendages used for locomotion.

On one of the rear walls is a fossilized giant bulldog fish, found in the chalk beds of Kansas. Scientifically known as *Portheus molussus,* this huge creature is 15 feet, 8 inches long. When this Teleost fish lived, Kansas was submerged under a shallow inland sea, somewhat like the Mediterranean. Eighty millions of years ago this prehistoric inland sea swarmed with giant sea-lizards, huge marine turtles and great and small fishes of many varieties, but the giant bulldog fish must have struck terror into the hearts of his contemporaries.

There is a "fossil aquarium" in the Hall of Fossil Fishes. Here has been reproduced a bit of a sea bottom that covered Scotland about 300 million years ago. The exhibit shows some of the common fish of that era, including "spiny" sharks and "joint-necks." Some of the fishes just beginning to appear at that time are shown also in this aquarium group. These relative newcomers, ancient ancestors of our modern bony fish, still look very primitive.

The most interesting of this group, are the "lobe-fins." Back in those early days some of these lobe-finned fishes, because of their possession of primitive lungs in the form of moist internal sacs into which they could suck air, were able to breathe out of water. So, filled with the spirit of adventure, they finally crawled up out of the water by means of their powerful paired fins upon which they could walk in a crude but effective way, dragging their long bodies on the ground. Thus began the vertebrate peopling of the dry earth, for they are the ancestors of all of the land vertebrates.

Also shown in the "fossil aquarium" are some early Ganoid fishes. The present day survivors of this group are the sturgeons, garpikes and bowfins which can be seen in the Hall of Fishes on the first floor in a series of habitat groups. The true Ganoids were the forerunners of the hordes of Teleosts or bony fishes that swarm the oceans today.

Through technology some of the great arid plains of the West have been reclaimed. Fields of green, orderly rows of orchard trees, and meadows of flowers have been brought to life by man's use of irrigation. But no human ingenuity will ever be able to reproduce the West of the Mesozoic Era, the Age of the Reptiles. This great age began about 200 million years ago and lasted until approximately 70 million years ago, covering a span of almost 130 million years. During much of the Mesozoic the now bare and arid plains were lands of lakes, rivers, swamps and luxuriant vegetation.

PROBING THE PAST

Many of the creatures which flourished in this golden age of amphibians and reptiles have finally come to rest in the American Museum of Natural History's fossil collections and the outstanding specimens are on view in the paleontology halls on the fourth floor.

Up to the Mesozoic Era, the predominant form of vertebrate life was represented by the fishes, and up to that time life was limited to a great extent to a watery environment. However when the lobe-fins and the lung-fishes heralded the beginnings of land animals, some prehistoric fish literally took the first big step in the evolution of land vertebrates and another of the milestones on the pageant of life was passed. Thus, some 300 million years ago the first land-living amphibians came into being. The die had been cast and although for a while these prehistoric ancestors of our modern salamanders, frogs and toads held sway as the highest form of animal life to inhabit the planet, they were soon to lose their rank to their relatives the reptiles. Of all the known prehistoric reptiles, the most amazing and fantastic are the dinosaurs, the chief of prehistoric creatures to many people, though there were others of greater importance in the story of vertebrate evolution.

The Museum's collection of dinosaurs and other fossil vertebrates, now the finest in the world, had to be built practically from scratch. Its actual growth dates back to 1891 when the Department of Vertebrate Paleontology was started. Even before then, there were two American paleontologists Edward Drinker Cope and Othniel Charles Marsh, whose studies and collections were to have a great influence upon the Museum's work in this field. The fact that they were rivals whose feud drove them in a bitter race to outdo each other in collecting fossils in the Great Plains and the basins of the Rocky Mountains, played no small part in the success of their efforts. They combed the fossil beds of the West from 1870 to 1895. Both were wealthy and they spared no expense in outfitting numerous collecting parties. Today we know that the West of North America contains abundant fossils. Cope and Marsh acted as though they were precious and rare jewels, and engaged in keen and often cut-throat competition to secure the largest collection. They also raced each other in describing their finds and as a result published a great deal of material. Thus two great fossil collections were built in America. In both of these collections were many dinosaur specimens.

Today, most of Cope's collection rests in the American Museum of Natural History, for which it was bought by Morris Ketchum Jesup. The collection of Cope's rival, Marsh, reposes in the United States National Museum in Washington, D.C. Thus two institutions, which might be considered friendly rivals, profited by the unfriendly rivalry of Cope and Marsh.

Professor Henry Fairfield Osborn along with his friend William Berryman Scott were fellow students at Princeton University. As early as 1877, Osborn, Scott, and another young man named Francis Spier, Jr., had organized and completed, while still students at Princeton, a collecting expedition to the Bridger Basin in Wyoming to gather Eocene Mammals. Both Osborn and Scott were disciples of Cope and as evolution was the burning topic of the day, Osborn had travelled abroad to study under the great embryologist Balfour and later under T. H. Huxley.

When Osborn came to the American Museum, his first

task was to assemble a staff of paleontologists and preparators that would enable the Museum to begin an active program in the field of vertebrate fossils. This great team was to include Cope's assistant, Dr. Jacob Wortman, the gifted paleontologist, Dr. William Diller Matthew, Dr. Walter Granger (one of the ablest of collectors whose work in the Gobi Desert as second in command to Roy Chapman Andrews of the Museum's ambitious and productive Central Asiatic Expeditions was but a small part of his contribution to paleontology), Dr. Barnum Brown who devoted a lifetime to the collecting and studying of dinosaurs, and Dr. George G. Simpson, who now heads the Department and whose studies in evolution and in the field of fossil mammals are outstanding. Important to the task are the services of preparators and field collectors and the efforts of Dr. Osborn added to the distinguished staff of scientists such skilled and talented men as Mr. Adam Hermann, who had been trained by Dr. Marsh, and Albert Thomson, Charles Lang and Otto Falkenbach. These men and their successors in the fossil laboratory at the Museum are the ones whose patience, skill and mechanical artistry have produced the lifelike restorations and exceptionally fine mounted fossil skeletons that make the fossil exhibit at the Museum so outstanding.

When Osborn, who was professor of comparative anatomy at Princeton, came to New York he became De Costa professor of zoology of Columbia University where he was the leading individual in the organization of the University's new Department of Zoology. In connection with this latter activity, some of Osborn's graduate students later became key figures on the Columbia faculty in comparative anatomy, paleontology and neurology. These included Doctors

Gary N. Calkins, J. H. McGregor, Henry E. Crampton and William King Gregory, who was to serve many years on the staff of the Museum as its Curator of Fishes and its Curator of Comparative Anatomy. The cooperative relationship between Columbia University and the Museum continues to flourish to this day.

According to Dr. Edwin H. Colbert, the contemporary Curator of Fossil Reptiles and Amphibians at the Museum, there is no mystery about the process of hunting and finding fossils. "It is," says Colbert, in his *Dinosaur Book* "mainly a combination of good geological judgment, horse sense, perseverance and hard work." The fossil hunter must first know where to look for fossils. If dinosaurs lived in the Mesozoic Era then their remains are to be found in rocks laid down in that time and nowhere else. The word fossil derives from its Latin original, *fossilis,* which means that which is dug up. A fossil digger must not only know where to dig but he must know what a fossil looks like. Sometimes just the slightest bit of what he seeks may be protruding from the sandstone cliff which covers the rest of the remains, or a broken bit may have tumbled slowly down to the bottom of the cliff and may rest there mingled with rock debris. A skilled fossil hunter knows his sedimentary rock and learns with experience in the field to spot the telltale traces of what may be a rich fossil bed.

Nature preserves fossils in many different ways. Most often the fossil consists of a replacement in stone or mineral matter of the original organic material. Before this replacement can occur, the creature's remains must have been protected from the elements and the process of decay by being shielded or covered up soon after death. This happens to creatures whose dead bodies are covered up by the ooze at the ocean bottom or in the mud or sand at the bottom of a lake or river, or by the wind-blown sands of a desert, or for that matter, in natural asphalt as in the famous tar pits of Rancho La Brea near Los Angeles. Once covered and protected, a slow percolation of mineral matter which is suspended in water begins to fill all the pores and air spaces in the hard parts of the carcass until gradually the original organic matter is replaced by mineral matter. This does not happen in all cases, many fossils are the actual preserved bones, shells, teeth or hard parts of the animals or plants that died in prehistoric times. These, however, are usually organisms that have lived in fairly recent times.

Often fossils are found as molds or casts. This occurs when the original organism has decayed or dissolved but leaves a hollow impression in the sediment which encloses it. Later this hollow is filled by mineral matter thus producing a cast of the original. Fossils are not necessarily parts of what was once a living organism. They may be traces of the organism such as burrowings, tracks, nests and even the marks made by leaves brushing across mud.

Finding the specimen in the field and getting it to the Museum is but one part of the story. Unpacking and removing the fossil from its rock matrix and restoring or assembling its parts are long and arduous tasks the story of which will be told later in this book, in the chapter "Behind The Scenes."

The vertebrate fossils of the Museum's exhibition collection are shown in six exhibition halls on the fourth floor and in the corridors connecting these halls. Besides the Hall of Fossil Fishes in the Southwest Tower there are the Jurassic

Hall, the Cretaceous Hall, the Hall of the Age of Mammals and the Hall of the Age of Man. In addition a new installation is under way which will contain material bridging the gap between the Age of Reptiles and the Age of Mammals. Eventually it is hoped to have all the fossil halls so located that the visitor can start with the invertebrate material and, in a normal progression, pass exhibits which will take him through the successive ages of fishes, amphibians, reptiles, mammals, and man. At present this sequence is broken because the fossil invertebrate and the fossil fish collections are out of sequence. However the visitor gains a semblance of this chronological squence when walking from the fossil amphibian exhibit at the far end of the Jurassic Hall, through the Cretaceous Hall, into the Hall of the Age of Mammals, and finally the Age of Man Hall.

Many of the fossil amphibians in the Museum's exhibit resemble some of the larger salamanders of today but in striking contrast so far as size goes are the *Eryops* and the *Diplovertebron*. The larger and more robust *Eryops* was the dominant Permian form. The fossils of this six-foot crawler were found in Texas. Close by is *Seymouria,* the "grandfather" reptile. Not greatly different in appearance from *Eryops, Seymouria* formed an intermediate link between the Permian amphibians and the reptiles that were to follow.

Related to *Seymouria* are the large pariasaurs, typified by *Scutosaurus* from the Permian beds of Russia and South Africa which shows an early tendency to giantism in the reptiles. It is as large as a good-sized cow but is very primitive, with a small head, clumsy, squat body and short legs that bow outward.

The Museum's collections abound in other prehistoric

reptiles besides the dinosaurs. Great phytosaurs, which resembled modern crocodiles, giant turtles, and an amazing assortment of flying reptiles like the *Pterodactyls* and the giant *Pteranodon,* mounted to show the great expanse of its outstretched wings, are to be found in the Cretaceous Hall and in the corridor between the two dinosaur halls. The Fort Lee *Phytosaur* whose remains were found in the red sandstone of the Newark series which underlies the picturesque Palisades of the Hudson at Fort Lee, New Jersey, is the fossil reptile found nearest to the location of the Museum itself. The Giant Fossil Tortoise of India, weighing over a ton, dwarfs its modern cousins, the big Galapogos tortoises. The flying reptiles found in the chalk and lithographic limestones of Bavaria range in size from delicate pterodactyls no larger than a swallow, to the large *Pteranodons* with wingspreads up to twenty feet. A mural showing a restoration of a Jurassic landscape shows the flying reptiles, which really were gliders and not flappers like birds, crawling up a prehistoric cliff and launching themselves into long glides over the sea, dipping down now and then to catch fish with their long, beak-like jaws.

Dinosaurs were cold-blooded animals, related to their modern reptile survivors, the snakes, turtles, lizards, and crocodiles. The name *Dinosaur,* which means "terrible lizard," was coined by Sir Richard Owen of England, the man readers will remember as the British Museum's Superintendent who was so encouraging to young Albert Bickmore. Owen was the man who established the science of vertebrate paleontology in England. One of the first dinosaurs to be found was that discovered in England by Dr. Gideon Mantell. In 1825 he published a paper on his discovery which he called the

Iguanodon. Mantell's house is still standing in Lewes, south of London and on a brass plate fastened outside the door are inscribed the words, "He discovered the Iguanodon."

Another early digger of fossils was, strange to say, not a learned paleontologist but a little twelve-year-old English miss named Mary Anning. Back in 1811 she discovered the first skeleton of the great marine reptile *Ichthyosaurus* while hunting for fossil shells. Encouraged by this early success Mary Anning continued her interest in fossil collecting and subsequently discovered the first *Plesiosaurus* skeleton in 1821 and the first *Pterosaur* or flying reptile in 1828. The Museum has fine specimens of the great marine reptiles *Icthyosaurus,* and the long-necked *Plesiosaurus* which are in the corridors outside the Jurassic Hall along with the fossil of another great marine lizard, *Mosasaur.*

The first dinosaur to be dug up in North America was found in the town of Haddonfield, New Jersey, a suburb of Philadelphia. In 1858, this fossil was given to the Philadelphia Academy of Natural Sciences by W. Parker Foulke. It is the skeleton of one of the duck-billed dinosaurs, *Hadrosaurus.*

Not all dinosaurs were large by any means. Some of the older forms ancestral to the larger types were about the size of turkeys. Recently a Museum expedition to the Southwest found some remains of a species not much larger than a good-sized jack rabbit. Dinosaurs have been found on every one of the major continents, in fact it seems evident that these creatures had a range which was as wide as the earth itself. Since the American Museum field parties did most of their digging in the American West, much of the Museum's dinosaur collection is from our own country.

Brontosaurus, the Thunder Reptile, dominates the Jurassic Hall. The skeleton of this mighty reptile is 67 feet long. It stands 15 feet high at the hips, the highest region of its back. In life, *Brontosaurus* must have weighed almost 35 tons, almost 7 times as much as the largest African Elephant on exhibit in the Akeley Hall of the Museum. *Brontosaurus* belongs to a sub-order of the saurichian dinosaurs known as the sauropods. These sauropods who lived late in the Jurassic and in Cretaceous times were the largest of their kind. Some were larger than *Brontosaurus*. The partial skeleton of *Diplodocus* in a wall case near the huge *Brontosaurus* is over 70 feet long and some specimens have been found of other sauropods that have exceeded 80 feet in length.

At this time it might be well to remark that the largest of all animals to inhabit the planet may still be alive. Dinosaurs, large as they were, are a poor second in size to the bulkier and longer blue whales which inhabit the oceans today. Whales have been caught which have been over 105 feet in length and have weighed almost 100 tons. Thus the sauropods with their long slender necks and their extremely long tapering tails were less than one-third the bulk and weight of our large contemporary whales.

But the largest of the dinosaurs probably were the greatest land animals that ever lived. The whales are of course buoyed up by water and even the largest of the dinosaurs probably spent much of their lives half submerged in the marshes where they fed upon the lush plant life. The head of *Brontosaurus* was small, tiny in comparison with its bulky body. In order to keep so large a body properly nourished this great creature must have spent all of its waking moments eating. It had rather small, peg-like teeth, and these only in the front of the jaws. The nostrils were located on the top of the skull, another indication of the animal's adaptation to a watery environment. It probably found protection from the smaller carnivorous dinosaurs of its day by means of its great size and also by its apparent ability to wade or swim out to deep water where only the top of its head would protrude above the surface.

Once there was an old lady who upon seeing the dinosaurs expressed her wonder at how the scientists who discovered these creatures were able also to discover their names. Since these animals lived long before man was on the scene to name them, in life they obviously had no name at all, a deficiency which has been remedied by the scientists whose nomenclature has become standardized in a universal language of science. Based on Greek and Latin roots it furnishes a simple way to describe the shapes and characteristics of the animals. The binomial Linnean system, whereby the first name indicates the genera and the second the species, and which one sees attached, usually in small print, on all the museum specimens that are labelled, has served its purpose well even though it may confuse the layman who fails to grasp its function. Take the horse for instance. It is *pferd* to the German, *cheval* to the Frenchman, *caballo* to the Spaniard and countless other names for the folks who speak other languages. *Equus caballus,* the Linnean term for horse means the same to all scientists, no matter what their native tongue.

Stegosaurus is a dinosaur that has been named from the Greek word *stegein* meaning cover. This "cover" lizard (for the suffix *saur* means lizard in Greek) is thus named because of the armor which covers its back. When *Stegosaurus* was first exhibited at the Museum, the overenthusiastic gentlemen of the press spoke of it as the dinosaur with two brains. Actually it has a tiny head which once housed a very small brain. In dinosaurs of this type there was an enlargement of the spinal cord, some twenty times the size of the brain, in the region of the hip. However, this was not a second brain, but functioned as the motor control of the large hind legs and the tail. In addition to the plates of bony armor along the spine, the *Stegosaurus'* tail was equipped with four large spikes. These spikes were weapons and were probably most effective when the animal lashed at its enemies with its tail.

The most ferocious appearing dinosaurs in the collection are quite naturally the large carnivorous types like *Allosaurus* and the giant *Tyrananosaurus rex*. The Museum's mounted *Allosaurus* specimen is posed to represent such a creature feeding upon the carcass of a *Brontosaurus*. *Allosaurus*, which thrived in the Jurassic, and its larger cousin, *Tyrannosaurus*, which dominated the Cretaceous period, are theropods, bipedal meat-eaters whose bodies were pivoted at the hip and whose long legs, ending in claws not unlike those of a bird, must have enabled them to cover the ground rapidly with giant strides. Their tails served as counterbalances for the weight of their upper bodies and heads. The teeth of these beasts were long and blade-shaped and were set in large jaws capable of a great gape or bite.

Tyrannosaurus, the "King of the tyrant lizards," was one of the most destructive land creatures ever to have trod upon the earth. The Museum specimen is 47 feet long and the top of its head reaches 18½ feet above the ground. In life, this monster must have weighed between six and eight tons. The head on the mounted skeleton is a plaster cast as the real head has been placed on exhibit separately where the visitor can examine it closely without having to resort to a stepladder.

Many other Jurassic forms are shown including the horned dinosaur, *Triceratops*, an assortment of duck-billed dinosaurs including several *Trachodons*, and a fossil mummy, that of part of a *Trachodon* which shows the pattern of the tough skin which looked somewhat like pebble-grained leather. Some of the ostrich-like dinosaurs are also exhibited—the Theropods such as *Struthiomimus*, and the "bone-head" of all times, the *Pachycephalosaurus*. These "bone-headed" creatures had an inordinate thickening of the skull roof forming a heavy, solid covering of bone above the brain.

Near the fossil exhibits are several of Charles Knight's excellent oil paintings, showing some of the more prominent of the dinosaurs as they may have appeared when alive. More of Mr. Knight's paintings are in the Hall of the Age of Man and in the Chicago Museum of Natural History. Knight is probably the outstanding artist of the prehistoric and his carefully prepared illustrations of prehistoric creatures, checked for accuracy by leading paleontologists, grace many museum exhibits and are to be found in many popular as well as scholarly works on paleontology.

In 1900 Professor Osborn made the brilliant prediction that Asia would prove to be a great, if not the greatest dispersal center of prehistoric life. In the early 1920's it was planned to send an expedition to Central Asia to search for fossils of some of the earliest forms of animal and human life. It was recognized at that time that the fossil history of Asia was almost completely unknown. "Dragon's teeth," or fossil teeth and bones, had been found in China for some years before this, usually powdered and sold as medicine by Chinese apothecaries. Some interesting finds had been made in Java, Persia, and Asiatic Russia, but it was no easy task trying to raise money to finance an expedition to the Gobi Desert of Mongolia which most people were then convinced was a wasteland hardly worth investigating.

The expedition, with Dr. Roy Chapman Andrews as zoologist and leader, Dr. Walter Granger as chief paleontologist and second in command, had a superb staff of experts consisting of Dr. Charles Berkey of Columbia University as chief geologist, Clifford Pope as herpetologist, Dr. Nels C. Nelson as archaeologist, Ralph Chaney as paleobotanist. Dr. A. Z. Garber went along as surgeon. There was also a staff of topographers and photographers. This first Central Asiatic Expedition, in 1921 and the others that were to follow it year after year in the twenties, were among the most ambitious of all the museums' expeditionary undertakings.

The story of the way this expedition conquered the desert will be told in the chapter on Expeditions. It is mentioned now because of one of its finds, which, while just one item in the thousands of scientifically valuable specimens brought back from the Gobi, succeeded in capturing the public's fancy. The find was that of the Museum's famous dinosaur eggs.

One afternoon in the summer of 1923 the expedition made camp at the base of a spectacular series of sandstone bluffs which were known as the Flaming Cliffs. It had been a fairly strenuous drive to this point and camp was made at three o'clock in the afternoon, the cooks were given instructions to prepare some dried apple pies for dinner, and a vacation was declared for the remainder of the day. Enthusiastic fossil hunters, such as those on this expedition, knew just what to do with such a vacation. In a few minutes they had all wandered off down the bluff and soon were scattered among the ravines looking for fossils.

In less than an hour one of the men came running back into camp looking for his tool box. He reported the discovery of a dinosaur skull. He was soon followed by others who also had found fossils, and by evening every man had begun to dig out his finds. There seemed to be quite a few skulls of a small, horned dinosaur of the type later called *Protoceratops andrewsi*.

On the second day at the cliffs, when the men came in for lunch, one of the field collectors, George Olsen, reported that he was sure he had found fossil eggs. There was a great deal of joking but after lunch the men were curious enough to go with Olsen to his discovery. Then, as Dr. Andrews later wrote, "Their indifference suddenly evaporated for they realized that they were looking at the first dinosaur eggs ever seen by a human being." At first the men could hardly believe their eyes and even then they tried to account for the eggs in every possible way as an inorganic geological phenomenon. But close examination removed any doubt concerning their nature.

Now the fact that dinosaurs laid eggs was not too startling. Some contemporary reptiles produce offspring by laying eggs—others by living births. But never before had an expedition found the eggs of dinosaurs. The eggs could hardly be birds' eggs as there were no birds known from the lower Cretaceous, the geological formation in which the eggs were found. There were prehistoric birds not long after this but they were all too small to have laid eggs the size of the ones found. Then too, these were reptilian in shape, elongate, and not sharply pointed at one end as most birds' eggs are.

Three of the eggs were first exposed to view and other shell fragments could be seen partially embedded in the rock near by. The preservation of the eggs soon uncovered by the excited yet careful fossil hunters was exceptional. Some were crushed but the pebbled surface of the shells was as perfect as though the eggs had been laid that year and not 90 million years before. The shells were about one-sixteenth of an inch in thickness and the eggs were from five to seven inches long. Fine sand had filtered into breaks in the shells of the whole eggs and these were filled with solid sandstone. More eggs were found a few days later—twenty-five, altogether. Nearby were skulls and bones of the *Protoceratops*, the small horned dinosaur which it is thought had laid the eggs. A fine series of *Protoceratops* were found, from the egg stage, through very young and intermediate specimens to full-sized adults almost nine feet long. These dinosaurs and the famous eggs have long been one of the high spots of the fossil collection for the Museum's visitors.

Credit for the first find, now identified as that of a fragment of egg shell of the *Protoceratops* really should go to Dr. Granger who found it at Shabarakh Usu in 1922. Olsen, however, discovered the first nest and really enjoyed the thrill of discovery for it was not known definitely at the time that Granger's shell fragment was that of a dinosaur's egg.

The American Museum is extremely proud of its fine collections and exhibitions of early mammals. It is very fortunate to have been able to collect some outstanding series of such mammals as the extinct *Titanotheres*, the camel, horse and the elephants. It was fortunate that the Museum and the great American West were opened at about the same time for early field parties to the plains and plateau areas discovered many fine specimens of early American fauna.

Mammals are warm-blooded creatures, far advanced over their lower reptilian forbears in the matter of the development of the brain, heart, control of body temperature, higher metabolic rate and generally increased activity. Their greatest advance over the reptiles is in the matter of intelligence. Scampering around practically under the ponderous feet of the giant reptiles, whose dominant position they were to usurp, the early mammals, small in size, without armor, without large fangs, had but one great weapon to use to avoid becoming just another bite for some dinosaur. That weapon was the intelligence which went along with their larger and better developed brains.

The largest of the fossil mammals on display at the Museum is the great *Baluchitherium,* an ancient relative of the rhinoceros, which stood nineteen feet high at the shoulders and was thirty-four feet long. The Museum has the skull, the feet and some of the vertebrae on display, as well as a full-sized restoration in the form of a gigantic bas-relief on the wall.

The Osborn Hall of the Age of Mammals includes mammals chiefly of the Tertiary Period. The magnificent series of *Titanotheres,* hoofed mammals, related to the horses and rhinoceroses, contains specimens of this extinct creature which reveal it in various evolutionary stages from little creatures about the size of small dogs up to the huge bulky animals with bodies like that of rhinoceros but with flat shovel-like battering ram-horns.

Toward the latter part of the Tertiary Period there occurred the culmination of many lines of mammalian descent. It was as though Nature were experimenting on a vast scale trying to find the form of mammal best suited to meet the particular environment. As Dr. Colbert of the Museum's Paleontology Department has written, in *Natural History Magazine,* "These animals of the late Tertiary were for the most part wonderfully adapted to the environments in which they lived. They were suited to almost every conceivable 'ecological niche' of the late Tertiary world, that is, they were specialized to varied modes of life on the plain, and in the forest, to grass-eating and meat-eating, to burrowing and to climbing, to protection by virtue of their great size, and to escape by running. All the exigencies of life on the land were met by the evolutionary development in these animals."

Another exceptional evolutionary series in the Hall of the Age of Mammals is that showing the rise and development of the camels.

But the prize series of all is that which illustrates the evolution of the horse. When the Spanish Conquistadores came to Central America they were but a handful of men against the large armies of the Aztecs and their neighboring tribes. Yet these few Spaniards awed the natives with their steel armor, their firearms and, most of all, their horses. At the time of the Conquest there were no horses native to America. Yet, the fossil record shows that a remarkable evolutionary development of the horse took place in North America beginning millions of years in the past when the "dawn" horses, the tiny *Eohippus,* made their appearance. These little creatures, about the size of small dogs, had four toes on their front feet and three toes on their hind feet instead of hooves.

The supposed five-toed ancestor of the *Eohippus* is yet to

be found but the fossil collectors have succeeded in unearthing an extraordinary series of horses, each larger than the preceding, and each showing a step in the transformation of the pad-like four-toed foot to the hard one-toed foot ending in a hoof which characterizes the horse of today. The first *Eohippus* fossil was found by Dr. J. L. Wortman in the Wind River Badlands of Wyoming in 1880. It was purchased by the Museum in 1894. Later, in 1910, a Museum expedition uncovered another specimen and subsequently the Museum was able to acquire its world famous series by dint of its own field collecting and by purchase.

The late Professor Osborn told me the following amusing story in connection with the Evolution of the Horse exhibit. When Dr. Osborn was President of the Museum he acted as host to the Japanese Crown Prince who was then visiting the United States and accompanied the young Prince about the Museum. When they came to the Evolution of the Horse exhibit, he explained to the Prince that it was one of the prizes, if not *the* prize of all the Museum's treasures, and spent some little time discussing it. Then he told the Prince that he would be happy to present him with a plaster cast of *Eohippus* to take back to Japan as a memento of his visit to the American Museum of Natural History. The Prince, whose command of English may not have been of the best, and whose knowledge of vertebrate paleontology and the significance of the exhibit before him was even less secure, turned and thanked Professor Osborn and said that when he got back home he would be certain to show his countrymen the amazing way in which the Americans had developed the horse into a creature of considerable size from such a tiny beginning, and he hoped this would stimulate his people to get busy and do the same with their small Asiatic horses.

In addition to the fossil mammals already mentioned, the Osborn Hall contains exhibits showing the fossil ancestors of dogs, cats, and other living mammals as well as some excellent murals by Charles Knight depicting mammals of various periods in the Age of Mammals.

More recent fossil mammals are to be found in the adjoining Hall of the Age of Man. As these were contemporaries of early man, their story is best told along with that of early man. However, as a prelude to the story of man, it is well to mention some of the smaller fossil mammal specimens in the Hall of the Age of Mammals. Most visitors are inclined to pass these by because they are small and fragmentary. Yet, these are among the rarest and most interesting of fossils, those of the early primates (lemurs and monkeys, etc.). These unique specimens are known throughout the paleontological world because of the light they shed upon the earliest stages in the origin of man.

CHAPTER FOUR

Animals without Backbones

SOME OF the most beautiful creatures in all of nature cannot be seen by the naked eye. Were it not for the microscope a vast and dazzling parade of animal life would completely escape human vision. The American Museum of Natural History has worked long and hard to make it possible for all its visitors to discover and enjoy the delicately beautiful forms of these microscopic animals.

Perhaps the most conspicuous inhabitants of this universe of minute creatures are the teeming Protozoa, single-celled animals which constituted one of Nature's earliest experiments in animal life. The experiment has been successful for the protozoa still abound in all the zones of the earth where moisture is found, from the equator to the arctic regions and from the deepest ocean abysses to the highest mountain peaks. They swarm in the open ocean in uncounted billions and multiply unceasingly in ponds and streams. Their fossil shells form great deposits of rocks which, once under water, have been uplifted to form part of our continental masses. No habitat has been overlooked by their myriads. Some forms penetrate the tissues of other higher forms of animal life. They flourish in the intestinal tracts and blood streams of their hosts, including man. They are of tremendous significance to man since they can bring disease and death. We use their fossils when we use chalk. The famous White Cliffs of Dover are composed of limestone built up under water of their tiny limy shells. The oil geologist is interested in the fossils of protozoa and a whole science called Micropaleontology devotes itself to the study of tiny organisms that lived long ago.

The Museum has been able to show its visitors the beauties and wonders of this protozoan universe as the result of scientific team work. The scientists of the Museum's Invertebrate Department, and the skilled artists and modelers working in close cooperation with these scientists, have produced a series of exquisite colored glass and wax models of these microscopic organisms, some enlarged 1,000 diameters. Seen thus in three dimensions and accurately reproduced as to shape and proportion, these models rival the beautiful art of the most skilled jewelry designers.

The glass models used to be on exhibit in the Darwin Hall along with a great amount of other invertebrate material. Darwin Hall has now been dismantled and the exhibits that it contained have been rejuvenated and are on display in the gallery of the Hall of Ocean Life.

The models in glass are the masterpieces of Hermann Mueller, one of the Museum's finest craftsmen, who has now retired. Mueller's tools were as simple as his products are intricate and beautiful. He used a gas burner to heat the rods of glass, and a flat steel blade, sharpened along one side, to snip them into desired lengths. The rest he did with his deft fingers and his lips. Blowing glass is a fascinating spectacle, and many visitors behind the scenes often became so enthralled standing beside Mueller's workbench that they would forget the other activities they had come to see.

In 1870, the Museum received its first shell collections, from Mr. Albert Storer, Mr. Frank Daulte, Professor Bickmore and Mr. Coleman T. Robinson. Mr. Robinson's collection alone contained about 4,000 specimens representing almost 1,000 species. In 1881 the Department of Marine Zoology was established, with Dr. Joseph B. Holder as curator.

The Department of Conchology was firmly established in 1874 when Miss Catherine L. Wolfe purchased from Dr. John C. Day, and gave to the Museum as a memorial to her father, John David Wolfe, the Museum's first president, a fine collection of shells along with a rare conchological library. It contained 50,000 specimens representing 10,000 species. It was put on display in the Arsenal until the first permanent section of the Museum opened and then it was prominently displayed on the first floor.

In 1890 the Museum obtained the Steward Collection of beautiful and rare marine shells and the Crooke Collection of land and fresh-water shells, the latter containing 3,000 specimens and 2,300 species. Other early additions were the Haines Collection, purchased in 1895, and the Cope, Storer and Constable Collections, which expanded the total number of specimens to over 200,000 and the number of species to over 16,000 representing shells from all parts of the world. Year after year, additions were made until today individual shell specimens number almost ¾ of a million, representing over 35,000 species. The bulk of this collection is on public view in the gallery of Ocean Life Hall. Others are in cases below the exhibit cases or in the shell laboratory.

The Invertebrate Department, after having been incorporated into the Geology Department in 1888, separated once more in 1902 as the Department of Invertebrate Zoology, under Dr. William Morton Wheeler, has now been incorporated with the Department of Fishes and Aquatic Biology. Originally the Department of Marine Zoology included fishes, reptiles and amphibians as well as invertebrates. In 1909 insects and shells were incorporated with the Department of Invertebrate Zoology and in 1911, the separate departments of Ichthyology and Herpetology were set up. For some time insects and spiders have been the concern of a separate department which has in its exhibition and study collections almost 2,000,000 specimens. More than ¾ of these are insects, the remainder spiders.

The Museum's marine invertebrate exhibits now under the direction of John C. Armstrong, Assistant Curator consist of displays of single specimens and a number of large models along with many habitat groups which show these creatures in their natural environments. Two of these groups in the Ocean Life gallery show scenes magnified many times natural size. The *Bryozoa* Group represents two square inches of sea bottom as though enlarged under a microscope to an area five feet square. Looking through the front of the case, made to appear as a giant magnifying glass, the visitor can see marine plants magnified to tree-like proportions, encrusted with colonies of *Bryozoa,* composed of thousands of individuals, each of which builds a shell of vase-like form. Animals associated with the *Bryozoa* are shown including tubeworms, and sea spiders. Nearby is a companion exhibit, the Rotifer Group. Here a cubic half-inch of pond bottom has been enlarged one hundred diameters or cubically, one million times. Thus the visitor beholds the transformation of a minute area into a towering aquatic forest peopled by rotifers and other strange creatures normally invisible to the naked eye.

In the shell collection the largest specimen is that of *Tridacna,* the giant clam. The *Tridacna* on exhibit weighs 579 pounds. The shell of greatest value is the rare "Glory of the Sea," *Conus gloria-maris.* This is the rarest shell known and only about 12 specimens exist in all the collections of shells the world over.

The largest mollusk known is represented by a dramatic model of the giant squid *Architeuthis princeps* suspended above the balcony in the Ocean Life Hall. These giant mollusks reach a length of fifty feet. Two of the features of the Ocean Life Hall invertebrate exhibits are the large Coral Reef Group and its companion, the Pearl Divers Group.

The immense coral formation, weighing forty tons, once grew like some fantastic fairyland forest three fathoms below the surface of a crystal clear tropic ocean. Ripped from the center of a submarine forest of tinted limestone it was hauled to the beach, bleached, packed carefully in a bedding of pieces of sponge, and shipped a thousand miles to the American Museum of Natural History.

This largest of Museum groups was the brainchild and product of the ingenuity of Dr. Roy Waldo Miner and his talented associates in the Department of Living Invertebrates. Twelve years of hard work which included five expeditions to the Bahamas went into its procurement. Six tons of intricately placed steel framework support the forty tons of coral. The largest single coral in the group is a large, branching specimen twelve feet across and weighing two tons. Since the nature of the coral animal whose secretions build up the rock-like coral prohibits its being preserved in such a group, the actual living outer layer of coral animal had to be replaced in wax over each specimen. To insure accuracy, underwater colored photographs and motion pictures were made of the parent reef. In addition Museum artists lowered themselves into the clear waters and with diving helmets on their heads, set up weighted easels and painted their color notes on specially prepared canvas stretched over glass. One of the minor, irksome difficulties of this work was that the artists had to weight their brushes and every once in a while an improperly secured brush would float gently up to the surface twenty-five feet above the artist's head and have to be retrieved by an assistant in a rowboat.

The coral reef scene is in two parts. Water level is on a level with the gallery floor and the upper portion of the group rises from the gallery floor seventeen feet to the top of the group. The upper half of the group presents a vista of coral island, a quiet lagoon and a tropical sky. In the distance, the low-lying Andros Island of the Bahaman reef can be glimpsed with its fringe of cocoanut palms. At this level the visitor can look down, as over the side of a small boat moored right over the reef edge and see below him the splendid spectacle of the coral formations. Descending to the lower level, the visitor now finds himself in effect at the sea bottom, 17 feet below the surface and facing the reef in all its splendor of form and color, teeming with fish and invertebrate life. Rising to the surface are the branching trees of elkhorn coral. Off to the right in a cave formed by the coral are some blue parrot fish. In a clearing in the center are many of the rainbow-hued tropical fish, squirrel-fish, striped and gray grunts, blueheads, slippery dicks and spotted hinds. In the distance, a school of black angel fish swims past. Overhead a school of yellow tails can be glimpsed swimming around the coral cliff and nearby houndfishes seem to be darting about alarmed by the approach of the voracious barracuda. The delicate pastel shades of the coral and the sea-fans, sea-bushes and sea-plumes set off the more vivid blues, yellows, oranges and reds of the coral reef fish. Lurking in the crevices and caverns under the coral are a large rainbow parrot fish, Nassau groupers and a huge green moray. The two villains of the drama are the barracuda and the green moray. Both of these fish possess sharp teeth and are a hazard not only to other fish but also to any unwary swimmer or diver who may come within striking distance.

Built on the same grandiose scale as the Coral Reef Group the Pearl Divers Group represents a scene on the ocean floor near the coral atoll of Tongareva, or Penrhyn Island, a small ring-shaped island, eleven miles in diameter, located in the South Pacific Ocean about 2,000 miles due south of Hawaii. The scene at the Museum shows the bottom of a lagoon enclosed by the island and consists of three windows or panels. In the main panel two native divers are seen plunging into a gorge in the heart of one of the coral reefs which form the island. These men are gathering shells of the oyster *Melea-*

grina, the oyster that often contains the precious pearls. Their work is surrounded by danger. Dimly lurking in the background is a large shark and at the entrance of a nearby underwater cave slithers a good-sized octopus. One diver is at the bottom and is in the act of wresting some of the *Meleagrina* shells loose from their resting place. Near his right foot is another hazard, the beautiful, but poisonous sixteen-rayed sea-star, endowed with hundreds of poisonous sharp spines. Multi-colored fish of the South Pacific coral reefs swim past the divers. In the cave to the right shown in the second window or panel of the group are more octopuses who seem to have taken over this hiding place as their own.

There is another panel on the left side of the group which shows a cliff-like wall of the cavern where a colony of the man-trap clams is clustered. These *Tridacna* are not so large as the giant shell mentioned earlier, but if a diver got his hand or foot caught in one of these smaller ones he would be trapped, as the clams are firmly fixed on the coral.

A painting by F. L. Jacques above the Pearl Divers Group on the gallery level depicts the native divers in their outrigger canoes.

In the Insect Hall on the third floor of the Museum there are exhibits which treat of insect biology, including the relations between insects and vegetation, the importance of insects as carriers of disease, and general biological facts and theories concerning insects. There are a number of models of insects and of parts of insects. The center of the hall features a series of insect habitat groups. Several of these groups illustrate the life-histories of butterflies and others show the beneficial lady beetles as well as the most injurious Japanese beetles. Still other groups show some of the tropical insects such as the leaf-cutting ants and the stingless bees. The large model of the mosquito and of the house fly which used to be on exhibit in Darwin Hall are now in the Insect Hall. These are exceptional models, every fine detail of these harmful insects is carefully reproduced and the models are so large that the wingspread of the mosquito is about 36 inches across.

There is also a very useful exhibit showing how to collect and preserve insect specimens and numerous panels which display the beautiful moths and butterflies of the world. The exhibit of *Arachnida* or spiders is also in this hall. Of special interest to suburbanite visitors is "Insects in a Suburban Yard." This exhibit represents an attempt on the part of the late Dr. Frank E. Lutz to demonstrate to his Museum confreres, as well as to the public, the great number of insect species one can find in such a limited environment as a typical suburban backyard. Dr. Lutz collected every specimen in his backyard at Ramsey, New Jersey and when some other Museum scientist would boast of the number of species of some form of animal life collected in far away places Lutz would counter with his latest count of several hundred species collected right in his own backyard.

Starting a department of insects in a museum is like taking a bull by the tail. There are so many different kinds that no single institution appears able to afford a large enough scientific staff to care for the specimens properly. No one person can be an authority on all the kinds of insects. There are just too many of them. Thus the Museum's Insect Department under Dr. Mont A. Cazier is running a frantic race trying to keep track of its millions of specimens. By comparison with the staffs of some of the Museum departments dealing with vertebrates where the total number of species of any one group can be counted in thousands, the Insect Department staff faces the task of dealing with a form of animal life of which over one-half a million species have already been described and of which there may be as many as 4,000,000 or more kinds yet to be identified.

The first insect specimens came to the Museum in 1870 in the gift of the collection of Baron Osten-Sacken. In the same year Coleman T. Robinson presented his fine collection of moths and butterflies, containing 20,000 specimens and representing 3,000 species. Mr. Robinson was made curator of the collection and served generously in this capacity without pay until his death in 1872. Many collections have come to the Museum since then one of the largest of which is the Edwards Collection added in 1892. Henry Edwards was one of the most successful private collectors in the world at the time and his collection numbered about 200,000 specimens.

Many insects and spiders have been brought back by expeditions sent out by other departments in the Museum and the Insect and Spider Department has had its own staff out collecting in the United States, and in Central and South America. From time to time the department has experimented with exhibitions of live insect material, in spite of the difficult problem of feeding the living specimens. Some insects feed upon others and it is sometimes necessary to keep a stock of live insects on hand to provide food for the exhibition insects. For many years a working bee hive was located on the third floor in the Insect Hall. The hive was set next to a window and the Museum bees could come and go at will. The side of the hive was made of glass and visitors could watch the bees at work inside the hive.

THE WORLD OF NATURAL HISTORY

The Department of Animal Behavior of the American Museum devotes itself to research projects that deal with animals of all kinds. In recent years, Dr. T. C. Schneirla of this department has made a special study of the behavior of ants, particularly the army ants of the American tropics. In his laboratory is special equipment for studying the behavior of his tiny charges under various conditions. The equipment for these experiments includes a group of ant mazes, through which ants must travel in order to reach food and water. In some of the mazes there are three paths of different lengths all leading to the same goal. The object was to discover their ability to "learn" which of the routes was the shortest and also to discover whether or not any ants could retain the memory of which of the routes was the shortest in subsequent runs through the maze.

The insect exhibit also contains a large model of a fly in flight which shows the halteres which function somewhat as gyroscopes do and enable the fly to control and balance itself while going through its intricate flight patterns. Some of the visitors to the insect exhibit have been puzzled by the exhibit of the migrating monarch butterflies. They feel certain that someone at the Museum erred in placing altogether too many of these beautiful black and tan winged creatures on one tree. Actually the group is a faithful reproduction of an autumnal gathering of butterflies which assemble in great swarms at this time of year in the northeastern United States. When night comes great hordes crowd the leaves and branches of trees until there hardly seems room for one more. The flocks of monarch butterflies then move southward and the next spring individual females come north to reestablish the northern population.

ANIMALS WITHOUT BACKBONES

Few realize the importance of insects to mankind. Insects pollenize the flowers of many plants and trees, thus enabling these plants to bear fruit, produce seeds and thus perpetuate themselves, making the harvest of fruits and vegetables possible. In addition to this great contribution to man's welfare, insects make raw silk, and produce honey. Man has learned to obtain wax, shellac and certain dyes from insects. Many of the animals man uses for food or for other valuable purposes are plant eaters and thus are dependent upon the insects for making possible their supply of food. Then too, a great many birds, some mammals, frogs and their relatives and some small reptiles fancy an insect diet. The anteater lives almost entirely on termites. He is not alone in his taste for these tidbits as certain American Indians are also fond of eating termites. North Africans regard locusts as a delicacy while the Australian aborigines like insect grubs. The wise Chinese like to nibble on dried salted water beetles.

Certain insects are pests in every sense of the word; many are dangerous threats to man's very existence, but if man were to succeed in destroying all the insects in the world he would be making life most difficult and perhaps impossible. Just as man learns to control the harmful house fly, the malarial mosquito, the flea, the tsetse fly and many others, so he has also learned to appreciate the insects that help make life easier and more pleasant.

One of the major functions of any museum worthy of the name is education. Therefore, the exhibits in the Insect Hall try to tell both sides of the insect story. That is why there is an exhibit on the destructive cotton boll weevil and one on silk, one on the European corn borer and one on the honey bee, one on the cutworm and one on the lady beetle. One case presents examples of some of the products we use and enjoy which insects make possible. Many people are amazed to see the assortment of items in this case which includes cotton, silk, wool, fruits and vegetables of many kinds, coffee, tea, tobacco, chocolate, candy, and trout. Two items, trout and wool, seem out of place until one realizes that trout eat insects (remember the fisherman's flies which are made to resemble insects) and that sheep feed largely on clover which in turn happens to be one of the insect-pollenated plants.

CHAPTER FIVE

Scales, Feathers and Fur

SOME OF the biggest "fish" stories are true. Man's puny imaginative efforts are, all too often, put to shame by nature's own productions. Tales of mermaids, sea-serpents, and other fictitious inhabitants of the sea begin to seem mild when one learns more about the oddities in the fish world.

Imagine a female that weighs a thousand times more than its tiny mate who spends much of his life, hanging like a pendant, from the cheek of his supersized spouse. Such a creature does exist and can be seen in the American Museum of Natural History's Hall of Fishes. It is the deep-sea anglerfish. This same kind of creature actually fishes for its victims with a pole-like appendage which dangles in front of its head and lures its prey right up to its huge, gaping, tooth-lined jaws. Another oddity, the four-eyed fish can swim along with its two upper eyes scanning the surface while its two lower orbs search the depths.

It is not the purpose of the Museum's Department of Fishes to exhibit freaks of nature. The exhibit is planned to show fish life in its more typical phases and to display a representation of the more important families to which the 20,000 known species of fish belong. The subjects are classified and arranged according to their natural relationships with one another. Many are exhibited in habitat settings and their life histories, their importance in commerce, and their importance to the man who fishes for pleasure are indicated.

The American Museum had no exhibits on fish for the first thirty years of its existence. During this time it received occasional specimens but it was difficult to give these storage room, let alone find space for their exhibition. The early fish specimens in the collection consisted primarily of painted wax and plaster models which in some instances bore very little resemblance to their originals. Most of these specimens were in a small collection which had been purchased for about $700 from the Smithsonian Institution in 1886. Earlier, in 1876, the Museum of Comparative Zoology at Cambridge had given the Museum a series of fishes which had been collected by Louis Agassiz, and, from time to time, the United States Fish Commission sent specimens to the Museum.

For a while Fishes and Reptiles were under the care of the Department of Invertebrate Zoology, but in 1909, a new Department, that of Ichthyology and Herpetology was established, with Dr. Bashford Dean as Curator. Dr. Dean headed this department until 1925 after it had become restricted to fishes only, a separate Department of Herpetology having been established in 1920. After Dr. Dean's retirement from the Curatorship, the Fish Department was headed by Dr. William King Gregory who held that post until his retirement in 1944. Since that time the department has been modified and is now known as the Department of Fishes and Aquatic Biology and is under the direction of Dr. Charles M. Breder, Jr.

It was not until 1928 that the Hall of Fishes became a reality. Before that time the fish exhibits were scattered here and there in the Museum.

In addition to its public exhibition program the Museum's ichthyologists have carried out an extensive research and publication program. In addition to the studies of fossil fish, studies in fish osteology and the broader aspects of fish evolution have been pursued. Other researches have examined matters of taxonomy and the life history, embryology, behavior and distribution of fishes. A very important event in the ichthyological world took place in one of the offices of the Fish Department at the Museum in 1916 when the American Society of Ichthyologists and Herpetologists was established. The scientific journal "Copeia," now the official journal of that Society, was started by Mr. John T. Nichols of the Museum in 1913 and it was in his office that the "Ichths and Herps" got its start.

The interest of Mr. Michael Lerner, one of the Museum's present Trustees, in large oceanic fishes resulted eventually in the formation of the International Game Fish Association. This association concerns itself with the problems involved in game fishing on an international scale and acts as the record keeper for game fishermen the world around. Through its generosity the Association has provided the Fish Department with funds which have enabled it to conduct some important studies on big oceanic fish.

Both ends of the Hall of Fishes feature large spectacular

habitat groups, one end presents the Sailfish Group, the other a group of sharks sweeping down upon a helpless logger-head turtle. The sailfish exhibit shows a hooked Pacific sailfish clearing the water in a prodigious leap near the rocks of Cape San Lucas, Lower California. It is the gift of Mr. and Mrs. Keith Spalding and the fish in the group is actually a mounted skin of a sailfish and not a plaster cast as are some of the smaller fish in the hall. It is a fitting centerpiece for the fine collection of game fish at this end of the hall.

The nucleus of this collection of Marine Game Fishes consists of several ocean giants which were caught by the rod and reel of the late Zane Grey, known to millions as the author of many popular books on the West including the well-known *Riders of the Purple Sage*. To his friends at the Museum, Zane Grey was better known as a fine sportsman and the holder of many records for game fish than as an author. In addition to fish which he caught there are photographs of his prize catches. The largest fishes in this section were caught by Grey and by Michael Lerner. The heaviest single fish in the exhibit is a massive ocean sunfish *Mola mola* which weighed a ton and was caught by Zane Grey. This fish is a relative of the porcupinefish and its bulky body is enveloped in a rubbery skin which is more than two inches thick.

Mr. Lerner donated the Mako Shark, Tuna and Blue Marlin Groups to the Museum. The mako shark in the exhibit weighed 602 pounds when caught and is shown leaping completely out of the water in pursuit of mackerel. Mako sharks are among the swiftest of all sharks and are hard-fighting game fish. The large bluefin tuna was taken by Mr. Lerner off Wedgeport, Nova Scotia, one of the favorite fishing grounds for this fish for commercial as well as game fishermen. On another American Museum expedition to the waters of the Gulf Stream near Bimini in the Bahama Islands, Mr. Lerner hooked the big blue marlin seen in the Fish Hall group.

The fine exhibit on the life history of the swordfish, presented to the Museum by the International Game Fish Association, illustrates the life history of this remarkable game fish from the tiny egg to the huge, full-grown adult. These powerful creatures will sometimes turn back after they have been hooked or harpooned and charge the fisherman's boat. There is on record a case in which one drove its sword through 18½ inches of planking, 14½ inches of which were of oak.

Near the Game Fish exhibit is a habitat group which shows fish life of the Great Barrier Reef along the northeast coast of Queensland in Australia. The material for the Museum's group was collected off the shore of Heron Island at the southern end of the Barrier.

The largest of all the fishes is the whale shark. These harmless monsters have been known to reach a length of forty-five feet and it is believed that they may grow to lengths up to sixty feet. The Museum's whale shark is a young one and is shown swimming along the bottom of a pool off the shores of Bimini Island, the site of the Lerner Research station where much of the Museum's present studies in marine fauna originate. This small island at the northern end of the Bahamas is almost level with the ocean and it is well located within five miles of the Gulf Stream and a short distance from the deep submarine valley known as the "Tongue of the Ocean."

Whale sharks are found in the equatorial and South Atlantic regions, specimens having been caught off the coast of South Africa, the West Indies, Florida, Galapagos Islands in the Pacific and Cuba, but are seldom seen in northern waters. In Cuba they are called the *Pez Dana* because "Dana" is the Spanish word for the game of checkers, and this largest of the sharks, and of all fish, has a checker-board pattern on the dark, upper part of its body. In addition to white lines, the markings include a number of round spots which the Cubans say are the checkers. Each jaw contains about 3,000 teeth but an eighth of an inch long which are useless for biting. But after all, it doesn't really need teeth at all, for it feeds on small sea organisms which are caught in its gill-rakers as the water is strained through them, the gill-rakers functioning as does the whalebone in a baleen whale.

The largest of the truly dangerous sharks of today is the white, or man-eater shark. Normally an inhabitant of tropical waters, it frequently wanders north. The larger man-eaters are a dull, lead-white in color and sometimes grow to a length of forty feet. Their mouths are enormous and their teeth long and jagged, and while not as large as those of its extinct relative *Carcharodon* in the Fossil Fish Hall, are capable of great destruction. Their food generally consists of large fishes and sea turtles.

In addition to a fairly large mounted specimen of a man-eater in the Fish Hall, the Museum exhibit, "The Sea Rovers," shows a tremendous white shark, along with smaller sharks of other species, about to attack a logger-head turtle. The white shark is painted on the background of this group and one feels sure that if this were a living scene the white shark would get his full share of the prospective meal, since he so obviously dwarfs his companions.

The Museum's Fish Department bibliographer, Dr. Eugene Gudger, has spent a lifetime collecting records of fish behavior and from his files comes this story of the dreadful activity of a man-eater shark that swam into Northern waters in 1916. In July of that year a boy was attacked and killed by a shark while swimming in Matawan Creek, just inside Sandy Hook, New Jersey. A man who tried to go to the boy's rescue was also attacked, while still in shallow water, by the same shark. He died of his injuries. A few days before this there was a similar fatality at Spring Lake and another in Spring Haven. A short time later, on the 14th of July, a man-eater shark was captured suspiciously near Matawan Creek. Although this was only circumstantial evidence, it appears quite possible that this man-eater, so far away from his normal haunts, was responsible for the series of fatalities.

Lest this story frighten swimmers in Northern waters it might be remarked here that the chances of a swimmer being attacked by a dangerous shark is rare in these waters. Most of the nineteen or more different kinds of sharks that normally frequent our coasts are scavengers, feeding upon dead organisms, and not inclined to attack and feed upon living prey.

The sharks in the Museum's exhibits, like the lampreys, rays and lungfishes, are archaic fishes. They are very much like their prehistoric ancestors who swarmed in ancient seas and are a far cry from the biggest type of modern bony fishes, the Teleosts.

One of the strange-looking sharks in the Fish Hall is the hammerhead. Apparently evolved from some sort of blunt-headed fish, this creature occasionally reaches a length of twelve feet and is a swift swimmer. It can make very sharp turns by using its flattened head as a sort of bow rudder. The thresher shark has an enormously long tail and uses it effectively when rounding up schools of fish upon which it feeds.

The Blue Shark Group in the Fish Hall shows a blue shark and its newly born children swimming about under a ship in the Sargasso Sea searching for something to eat.

Skates and rays are much flattened-out relatives of the sharks. Like the latter, they have cartilaginous skeletons, and belong, with the sharks, to that primitive group of fishes known as the *Elasmobrancs*. The bodies of skates and rays are flattened and their pectoral fins have enlarged and widened into "wings." These "wings" are the chief means of locomotion as their tails have been reduced in thickness and serve as long, slender trailing rudders. Sting rays have whip-like tails bearing strong, saw-edged spines which can inflict most severe and irritating wounds.

The Museum also has a model of the manta or devil fish, a huge specimen of "winged shark" taken off the west coast of Florida that measures 17 feet across its widest expanse.

The unusual lungfishes are interesting, modern, relics of long past ages. Today they are found in Australia, South America and in Africa. They live in swampy regions which dry up partially each year. When the dry season approaches, the fish burrow deep into the mud. There they hibernate until the land is covered with enough water for them to emerge again. The Australian lungfish live in stagnant pools or waterholes and rise to the surface at infrequent intervals to empty their lungs and take in a supply of fresh air. One of the African lungfishes was dug up while still in the hibernating stage in the mud. Packed in a lump of mud, this remarkable fish was thus sent all the way to New York, where the hard, clay-like mud was chipped away and the fish released into a small aquarium tank. Here it proceeded to revive and swim about, apparently unharmed by its long and dry journey.

The "living fossils" or ganoid fish are a step higher than the Sharks and Rays, Lampreys and Lungfishes, but they still represent a very primitive kind of fish. Sharks and rays and the other primitive fish do not have scales like the modern fish. The ganoids have their scales covered by a thick, shiny, enameled layer whence came their name, as ganoid means "glistening." The living survivors of the ancient ganoids shown in habitat groups in the Fish Hall are the shovel-nosed sturgeon, the paddlefish of the lower Mississippi, the garpike, and the bowfin.

Teleost fishes which made their appearance during the Cretaceous period have become the dominant form of fish in our present day oceans. The majority of the specimens shown in the Fish Hall are of this type. All the common food fishes with which most of us are familiar, such as the salmon, the cod, halibut, mackerel, bass, trout, herring and tuna are teleosts. Some of the more unusual teleosts to be found in the Museum collection include the pipefish, the sea horses, flying fish, the silvery Moon fish or opah, the puffer, and the porcupinefish. The unusual shapes of these fish, their queer appendages and fins, are but modifications which have been evolved and which enable these creatures to meet the exigencies of their own particular environments.

Perhaps the most unusual fish in the ocean are those found in the extreme ocean depths. The deep-sea fishes, living thousands of fathoms below the surface of the ocean in total darkness and in cold water under great pressures, present a "rogues-gallery" of some of the most bizarre and terrifying visages ever produced by Nature. They must be brought to the surface by deep dredge nets, and were first seen in their native habitat only when Dr. William Beebe and Otis Barton penetrated the ocean depths in the Bathysphere, a hollow ball of steel, built to withstand the tremendous pressures of the lower reaches of the ocean, and equipped with portholes paned with windows of fused quartz three inches thick, which was lowered by a steel cable to a record descent of over 3,000 feet.

The American Museum's exhibit contains some of the actual specimens of deep-sea fish as well as a series of deep sea groups gathered in the ocean depths near the Galapagos Islands.

The vast oceans of the world remain as the great unexplored. We are just beginning to be able to penetrate the mysteries of the lower depths. Since the ocean has acted as a great conservator of life forms, and, since we can find ancient types of fish still living in its protective expanses, there is reason to hope that in years to come expanding frontiers of the sea will shed more light upon the fascinating story of the evolution of life. The ocean is a great source of nourishing food and the animal and plant life that abound in it are a great potential source of future food supply and contain the ingredients for many useful products. Thus advances in oceanography will benefit mankind throughout the world.

One of the exhibits in the Hall of Amphibians and Reptiles on the third floor recalls a dramatic story:

Like a monster from out of the pages of the past, a great lizard, almost 10 feet long, peered out from under the cover of jungle brush. It hesitated a moment as though to adjust its eyes to the bright light of the open clearing and then began to crawl slowly over the ground toward the object which had attracted its attention. Slowly, yard by yard, the beast came nearer and nearer, stopping every once in a while to raise itself up on its powerful forelegs to look around. Its defenseless prey was crouched in the tall grass, motionless in a desperate attempt to escape the great reptile's notice. This might be the description of a dramatic encounter between some prehistoric animal and its victim, but it happens to be a real-life situation which occurred in 1926 on a Museum expedition to Komodo, one of the lesser Sunda Islands in Malaysia. The figure crouched in the grass was Mrs. Douglas Burden, wife of a Museum Trustee who was the leader of this venture to secure specimens of the world's largest living lizard for the Museum's Hall of Amphibians and Reptiles.

In order to relieve the suspense it might be wise to continue the story of this tense and dangerous encounter in Mrs. Burden's own vivid words, in *Natural History Magazine:*

"Nearer and nearer he came, his grim head swinging heavily from side to side. I remembered all the fantastic stories I had heard of these creatures attacking both men and horses, and was in no wise reassured.

"My gun was propped against the blind where I had left

it a few moments before and was out of reach. Defosse [another member of the expedition] was out of sight and the reptile was continuing straight toward me.

"The creature was now less than five yards away, and its subtle reptilian smell was in my nostrils. Too late to leap from hiding,—if I did, he would surely spring upon me, rending me and devouring my remains as he had devoured the dead deer. Better to take my chances where I lay, so I closed my eyes and waited.

"Then I opened them in time to see Defosse's head appearing over the hill. The next instant there was a flash and a bullet buried itself in the great monster's neck. Like lightning he whirled and crashed toward the jungle, but the rifle once more did its work, and he lay still."

When the beast was measured a little later it was found to be almost ten feet long and weighed about 250 pounds. In its stomach was found the whole hind quarters of the deer it had recently devoured.

Not all of the specimens in the Hall of Amphibians and Reptiles were secured under such trying circumstances. The great Dragon Lizard Group shows these largest of all lizards in a typical setting on Komodo Island. The large male in the group has just killed a wild boar and, while he and a smaller female lizard are in the act of devouring their prey, still another giant lizard is seen dashing forth from the nearby jungle to dispute ownership of the dead boar with the first lizard.

The Museum's Amphibian and Reptile exhibit contains a series of habitat groups and displays of specimens that tell the story of the modern reptile and amphibian survivors of the creatures that once dominated the earth in the days of the dinosaurs. One year after the Museum's Charter was signed, the institution received its first collection of amphibians and reptiles. This was in 1870 when the collection of Prince Alexander Maximilian of Wied was acquired. The specimens in this collection had come from all over the world and later when expeditions went out to Mexico, Puerto Rico, Patagonia, Egypt and Siberia, more herpetological specimens were brought back. By the turn of the century, the Museum had several thousand. In 1905, a collection of almost all the species of amphibians then recognized in the United States and numbering 460 specimens was presented by Miss Mary Dickerson, later to be in charge of the Museum's herpetological department.

Today, the Museum's Department of Amphibians and Reptiles has a collection of over 50,000 amphibians and over 65,000 reptiles are listed in its catalogues. Its coverage is particularly strong in North and South American and African species, but its collections contain specimens from the world over.

The first land vertebrate was an amphibian relative of the large fossil *Eryops* seen in the Jurassic Hall. It was probably small, no larger than some of our present day salamanders. Certain conditions had to be met before the forerunners of the first amphibians could begin to be independent of the water from which they had come. They needed lungs in order to breath air, and they needed legs for walking about on land. Even their eyes, ears, and their sense of smell had to be modified for land existence. The first of these land creatures had moist skins like our salamanders and frogs. If they went too far from the water their skins would dry and they would die. Many modern amphibians have such skins and are just as much tied down to a moist environment as were their forbears. Snakes and lizards, on the other hand, have solved this problem of the moist skin. They evolved a thicker skin, the outer layers of which developed a horny substance. In time, snakes and most other reptiles developed a scaly skin. Now, no moisture was needed to keep their skin in a healthy state. The tougher skin and shell covering of the reptiles of today protects these animals, and often, as in the case of the armor-like covering of crocodiles and turtles, is a defense against other animals.

There are three principal types of contemporary amphibians, the salamanders, the frogs, and the toads while the four most common types of reptiles are snakes, lizards, alligators and crocodiles, and turtles. Amphibians lay their eggs in water, thus exhibiting another link with their watery past, while reptiles lay their eggs on land.

Exhibits in the Hall of Amphibians and Reptiles at the Museum show many variations of these creatures which have been brought about by the effort of certain species to adapt themselves to special environmental conditions. For example, one can see the way certain species of turtles have developed modifications of shape and skin texture as a result of living in or near water or in dry land burrows. Some of the lizards have sharp teeth useful for tearing and rending flesh while others have blunt, short teeth better fitted for cracking the shells of the mollusks and crustacea on which they feed. Tree frogs have developed sticky pads on their feet, which enable them to climb trees.

Snakes today crawl along by means of an intricate muscular movement which involves the scutes or bottom scales, and an exhibit in the hall shows how this is done. Yet, the snakes of today developed from ancestors which possessed limbs. Examination of the skeleton of the large python (one of which is on exhibit) reveals tiny legs which are hidden in the flesh during the snake's life and which are relics of the time when snakes walked like lizards.

Most of the poisonous snakes of the world are on display. Several habitat floor cases show copperheads, rattlesnakes of several varieties, including the largest, the big diamond back of the American Southwest, the cottonmouth moccasin, and the deadly and largest of all poisonous snakes, the king cobra of India. The cobra is shown confronting his fearless traditional enemy, the little mongoose.

Many of the snakes shown are harmless. Some, in fact, which eat harmful insects, and others, like the milk snake, which feed on rodents, are beneficial and merit our protection. In an attempt to dissuade the needless killing of beneficial reptiles there has been placed in the Museum Hall an exhibit which debunks some of the more prevalent snake yarns. Here one finds the true story of the milk snake, the "hoop" snake, the "whip snake" and other unjustly maligned serpents.

Exhibits in the hall also deal with the way some of the amphibians care for their young or at least for their eggs while they are hatching. One of the strangest methods shown is that of the Surinam toad of South America. As soon as the female lays her eggs her mate presses them into the soft skin on her back. A cover soon grows over each egg and as it develops it presses a deep pocket into the mother's skin. The young toads remain under their mother's skin until they develop into the adult form. Then they force open the skin covers and pop out to go about their business. Some amphibians such as the mud puppies never go through a complete change from egg to adult but spend their lives in the larval or intermediate stage.

On one side of the Hall, set off in a special corridor, is a series of ten habitat groups giving a picture of the home life of American reptiles and amphibians with the following subjects: "The Leatherback Tortoise," "The Giant Salamander," "The Bullfrog," "A New England Marshland in Spring," "West Indian Tree Frogs," "Reptiles of the Southwest," "Galapagos Iguana," "Rhinoceros Iguana," "Gila Monster," and "Florida Cypress Swamp."

The most powerful lizard in the Americas is the rhinoceros iguana (so named because it has a small protuberance resembling a rhinoceros horn upon its head) which frequents the desert regions of Santo Domingo. The Museum group shows a scene on the western shore of Lake Enriquilla, a dead sea in Santo Domingo which lies more than 130 feet below sea-level. This lizard lives in burrows which it digs in limestone banks. The females dig hills in the nearby sand flats and lay their eggs in these sandy hills in July. When the young iguanas hatch out as shown in the group, they frequently pull the bursting eggs with them to the surface. Like most other large iguanas, the rhinoceros iguana is a vegetarian and feeds on acacia beans, saona berries and other desert plants. The group shows a complete life history of these typical lizards from the egg to the adult.

Near the iguanas is the Gila Monster Group. The Gila monster, found in the Southwest, is the only poisonous lizard in the United States of America. There is only one other poisonous lizard known and that one is the Mexican Gila monster, a specimen of which is in the main part of the hall in a floor case. The Gila monster is about two feet long with a short heavy bulbous tail which almost appears to be another head. Its color is gray-black with irregular orange-rose patches and its skin has a pebbled and beady texture. It feeds on ants and birds' eggs and although poisonous, its bite is almost never fatal to human beings. This may well be because the lizard's poisonous fangs or teeth are well back in the lower jaw and it cannot inject the venom as can the rattlesnake, but must work it into the victim's blood stream by letting the poison seep into the wound made by the bite. The Museum Gila Monster Group shows a scene in one of the canyons of the Santa Catalina Mountains in Arizona. The group represents a sunset scene and in it also is a Sonoran racer, a snake of the area, and a desert tortoise looking for a hiding place for the night.

The Tree Frog Group shows a night scene in a West Indian forest in total darkness. A series of push buttons enable the visitor to light up the entire scene or any small part of it containing specimens.

One of the largest habitat groups in the Amphibian and Reptile Hall depicts a Florida Cypress Swamp in September. Numerous reptiles and amphibians find a home in these primeval swamps and the Museum group contains alligators, turtles, lizards, snakes, and frogs. A large alligator can be seen guarding her nest. Nearby young turtles are hatching from eggs which the mother turtle has hidden in the alligator's nest. The draping Spanish moss which hangs from the cypress trees in great profusion and the great tangle of vegetation along the floor of this jungle-like swamp half hide from view a multitude of creatures which share this interesting region with the alligators and turtles. Some of the breeding habits of turtles and toads are displayed and the group also illustrates the feeding habits of several species of snakes. It is a very good example of the profusion of reptile and amphibian life one finds in the warmer, moister tropical regions.

On a spring night in New England the marshlands are filled with sound. The bulk of this animal symphony is provided by the shrill peeping of the tiny tree frogs and the trilling notes by the gray tree frogs. The lower pitched notes are the mating calls of the male frogs and toads of other species which come to the marshes and ponds in the spring to breed. The Museum group named "A New England Marshland in Spring" shows dozens of these amphibian musicians hard at work calling their mates, their distended throats looking like small balloons as they call. The inflated throat sac acts as a resonating organ and accounts for the large volume of sound these relatively small animals are able to produce.

Near the spring marsh scene are the bullfrogs, engaged in all kinds of activity. One frog is in the act of swallowing a young water snake, and as though in eventual retribution, in another part of the group a large black snake is about to capture a bullfrog. An unusual sight for the visitor but a common occurrence in the life of the bullfrog, is shown in a skin swallowing act. As the frog sheds its old skin it disposes of this now useless bit of its anatomy by working it over its head and eating it.

In floor cases in the main part of the hall are additional exhibits of reptiles and amphibians. Here are the tiny flying "dragons," little lizards from the Malaysian region that have a loose fold of skin on each side of their bodies.

When they leap these spread out and act as "wings" enabling the lizards to stretch their jumps by gliding.

An exhibit of rattlesnake rattles features a mechanically activated rattle. The visitor sees a rattlesnake all coiled and ready to strike. By pressing a button at the front of the case the rattle can be made to shake as in life and the visitor hears the awesome buzz at close range. Nearby there is an exhibit concerned with snake bite. An enlarged model of the head of a poisonous snake is shown in which the poison sacs, the duct leading from the sac to the hollow fangs and the hypodermic-needle-like fangs are clearly visible. This exhibit also explains the remedies for treating the bites of poisonous snakes.

Tourists to the American Southwest have been thrilled by the Hopi Snake Dance ceremony, during which the dancers carry live rattlesnakes around in their mouths in the hope of bringing rain and insuring a good crop of corn. The dancers seem without fear as they handle the serpents with ease and even take hold of them in their mouths. The Indians use other snakes besides the poisonous rattlesnakes in these ceremonies as they go on a four day snake hunt prior to the dance and gather all the snakes they find. The mystery of how the Indians escape danger in the dance has been a baffling one and many explanations have been given but no definite proof had been found until recently to show that any of these explanations was correct. For example, it was thought that the dancers drank a liquid antidote before the ceremony or that the snakes were teased into striking at a bag or skin until they had expended their venom.

The answer to this mystery was finally obtained by Charles Bogert, the Museum's herpetologist. Another competent herpetologist, L. M. Klauber had witnessed a snake-dance but had not had the opportunity to examine any of the poisonous snakes used in the dance. The reason for this was that the Indians jealously guarded their reptilian charges and would not allow any outsider to handle the snakes. After the dance, the snakes were taken by the Indians and placed in snake shrines where they were supposed to carry back to the rain spirits the message that the tribe needed assistance. Bogert studied Klauber's report and decided that the only thing left to do was to secure one of the rattlesnakes used in the dance from one of the snake-shrines or piles of sandstone rock where the Indians would liberate them.

He attended the ceremony and was careful to avoid any actions which might arouse suspicion or antagonize the Indians. This was a serious religious ceremony for the Hopis and Bogert knew they would be very much upset by any interference. From a safe distance he watched the conclusion of the dance when the "gatherers" picked up great handfuls of the squirming reptiles and raced off into the desert to the snake-shrines. Through a small telescope Bogert saw them approach one of the shrines. He carefully noted its position and oriented it by means of some landmarks so that he could find it later. When the "priests" returned, Bogert went by a circuitous route to the shrine he had spotted with his telescope. When he arrived no snakes were in sight but he noticed a *paho* or prayer-stick stuck in the sand and saw that the ground nearby was sprinkled with the sacred cornmeal used in the ceremony. Just as he was about to start hunting for the snakes he realized he was being observed by a Hopi Indian standing about ten yards away.

Immediately Bogert pretended to be intensely interested in fragments of pottery which were on the ground and played at being an ethnologist until the Indian moved off. Bogert renewed his search and found a bull-snake. Pulling the bull-snake from out of the rocky crevice in which it had been hiding, he saw another snake in the crevice. This was a rattler and he soon bagged it. Rather than have the Indians learn the secret of the contents of the sack, he folded the bag lengthwise and managed to stuff it into the crown of his hat. Thus equipped with this addition to his headgear he was able to make his way back through the Indian village without causing any disturbance.

As soon as he arrived at a secluded spot he hastened to examine the mouth of the reptile, and found that *no fangs were present*. Still alive, the snake was forwarded to Dr. Klauber to examine. Klauber killed the reptile and preserved it according to Bogert's instructions and a close examination revealed an amazing fact. The functional fangs had been removed as Bogert had noticed when he first examined the reptile but in addition the embryo fangs (which, in life, will grow to replace fangs lost by accident) had been skillfully cut from the snake's jaw so that it had been made permanently and completely safe to handle. The poison sacs were present, but useless without fangs. Such a disadvantaged creature would eventually die from starvation as rattlesnakes are not constrictors and are able to kill the rodents upon which they feed only by injecting their venom into them.

This discovery indicated the skill and knowledge of reptilian anatomy which the Hopi "priests" possessed. Later examination of a poisonous snake used by one of the African snake-cults showed that these natives used the same technique.

The Department of Birds at the Museum was made a distinct entity in 1920 when it was separated from the Department of Mammals and placed under the charge of the late Dr. Frank M. Chapman. Today, this department, headed by Dr. Robert Cushman Murphy, can boast not only of having some of the outstanding bird exhibits to be found anywhere, but also of having in its vast study collections of almost a million specimens the world's outstanding and most nearly complete representation of avifauna outside of nature itself. The collection contains representatives of about 95 per cent of all the known species of birds. In addition to these specimens and the thousands in the bird exhibits, the Department of Birds also has a large collection of skeletal material, eggs and nests. These collections have been in the making ever since the American Museum of Natural History became a reality in 1869. They represent acquisitions by purchase, gift and field collecting projects of the Museum staff. Some of the important early acquisitions by purchase and gift include the Elliot, the Maximilian, the Vedray and Verreaux collections which were purchased in 1869 and 1870. The Smithsonian Institution presented a collection of birds' nests in 1874 and Dr. Daniel G. Elliot, who advised the Museum in its early purchases, presented 11,000 skins of North American birds in 1875. Dr. E. A. Mearns presented study collection material of bird skins and eggs from North America and Europe in 1882. In 1887 the Museum was able to exhibit its first series of bird groups made possible through the generosity of Mrs. Robert L. Stuart.

One of the most unusual of the Radiolaria, *Lithocircus magnificus*. These minute creatures construct a skeleton of natural glass from the silicon dissolved in sea water. *Lithocircus* has formed a vertical crystal hoop, with branching antlers extending in all directions. The central capsule houses the nucleus of this protozoan. A model in glass, many times natural size, this exquisite piece of museum craftsmanship is but one of many other similar wonders on view in the recently renovated exhibit of marine invertebrates in the American Museum's Hall of Ocean Life.

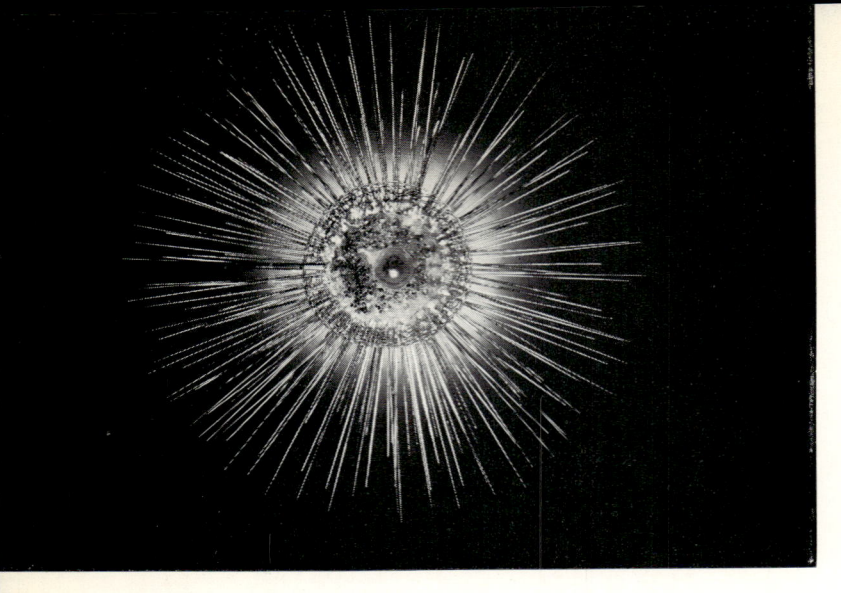

Above: In life, the beautiful *Aulonia hexagonia* is only ⅛ of an inch in diameter. The nucleus of this radiolarian is enclosed in an almost perfect sphere of glass lacework made up of multitudes of hexagonal meshes.

Below: Herman O. Mueller begins to construct one of his glass models of a tiny marine invertebrate by blowing a bubble from a tube of red-hot glass. Using very simple tools, the artist-craftsman shapes the glass to the desired form closely following the accurate drawings of the creature made after a microscopic examination of the tiny original.

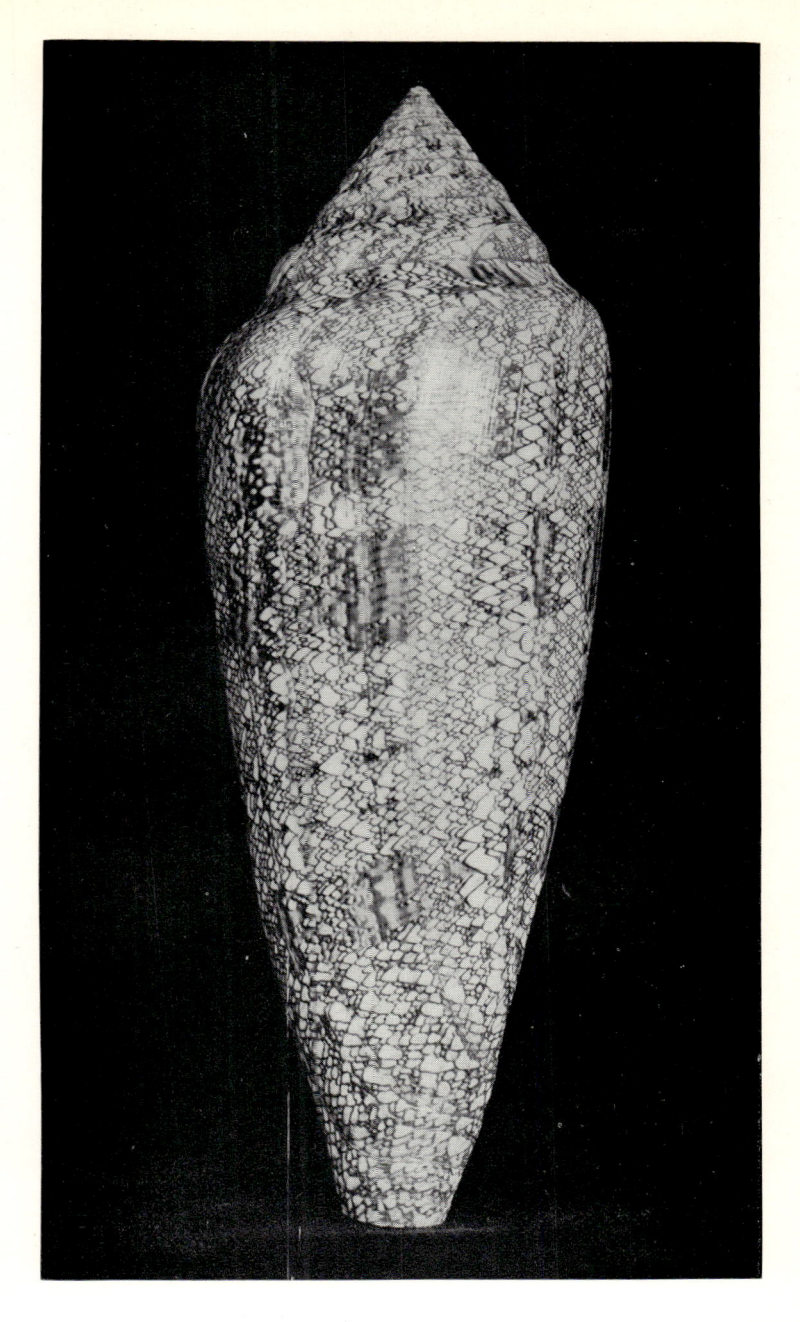

The Glory of the Sea (Conus gloria-maris), rarest of all the shells. Practically extinct, the last living specimen seen in 1838, no more than a dozen specimens are known to exist. The American Museum possesses this perfect specimen.

A detail from the Pearl Diver's Group in the Hall of Ocean Life. The scene shows a Tongarevan pearl diver plunging down into a coral gorge in a lagoon on the little South Pacific coral atoll of Tongareva. Ten tons of actual coral are featured in this exhibit. The prize of the specimens collected is the great *acropore* coral with its spiral turret shown in place in the lower right hand corner of this picture.

The divers in this group are engaged in gathering precious pearl shell despite the gruesome octopus, a poisonous sixteen-pointed sea star, armed with hundreds of sharp red spines, and a shark lurking in the distance. The group also includes many of the brilliantly-colored fishes which frequent this coral fairyland.

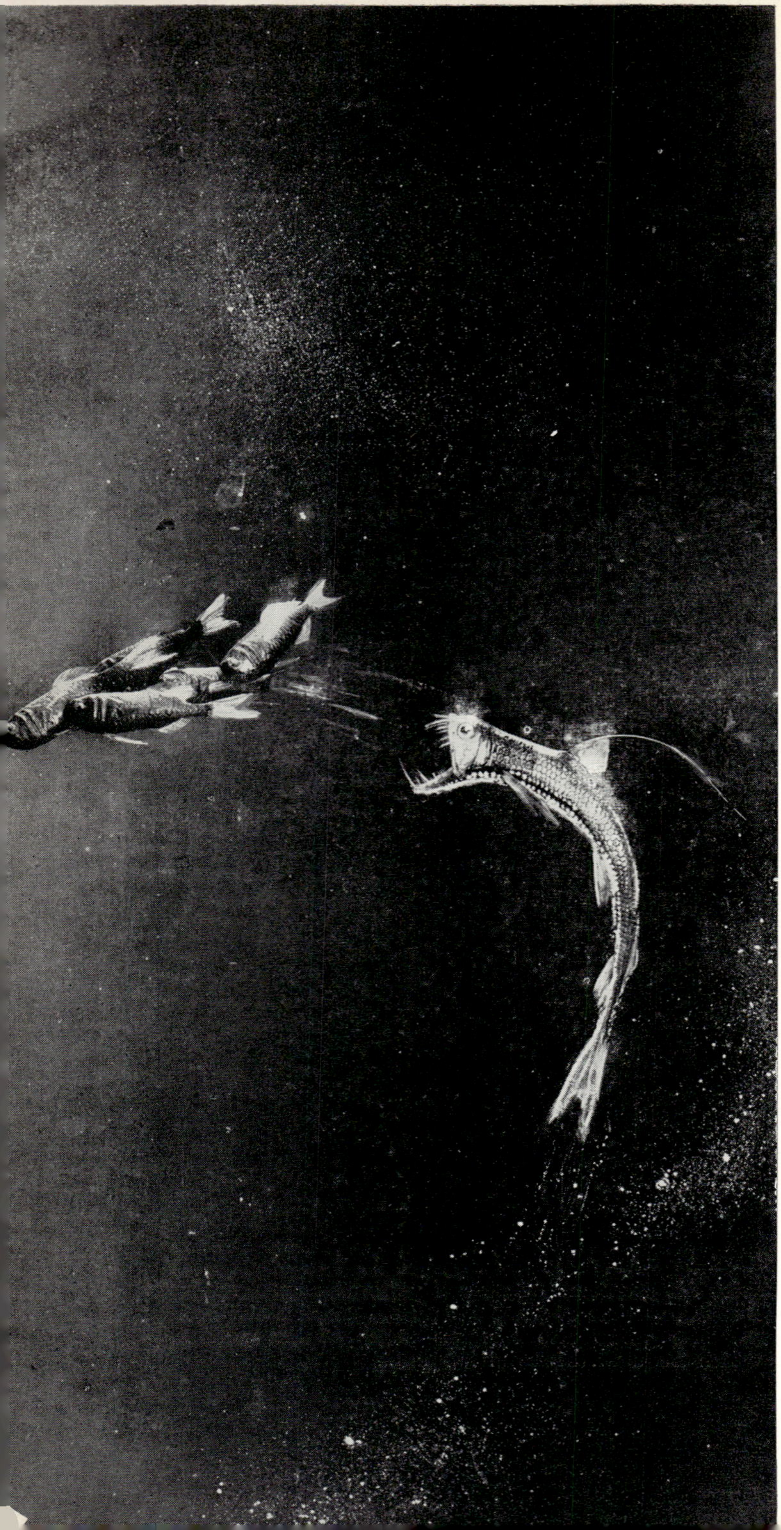

Above: "Trout's-eye View of a Fisherman." The studies of the Museum's Department of Experimental Biology resulted in exhibits like the one shown here. In this instance the image the fish sees is distorted because of the effect water has upon light passing through it.

Left: "The Dragon Strikes"—Half a mile below the surface a gleaming iridescent Dragonfish, *Chauliodus sloani,* pursues a small school of *Melamphids,* or Big-heads. These deep-sea fish are able to withstand the tremendous pressure and almost freezing temperatures of the great oceanic depths.

"A House Fly as Big as an Eagle." The Insect Hall contains this model of the common house fly enlarged many times natural size. Large models like this of small insects show the visitor many details otherwise never seen.

The largest lizards in the world are the "Dragon Lizards" from the island of Komodo in the Dutch East Indies shown in a Reptile Hall habitat group. Some of these monsters reach a length of 10 feet and a weight of 250 pounds.

The dodo of Lewis Caroll's immortal "Alice in Wonderland" was a real bird. This flightless bird became extinct in 1700 but once lived on the island of Mauritius off the coast of Madagascar. In 1907 The American Museum traded the bones of a Right whale for the bones of a dodo. This dodo skeleton and the feathered restoration shown above are in the Sanford Hall of the Biology of Birds.

Ship-Followers of the "Roaring Forties." Vast numbers of petrels and albatrosses spend their whole life, except during their breeding seasons, out in the open sea. These birds often follow sailing ships for food stirred up in the wake or tossed overboard. This habitat group in the Whitney Memorial Hall of Pacific Bird Life shows a scene southeast of New Zealand. The vessel seen in the distance is the Whitney Expedition schooner *France*.

Bird Life in the "land of little sticks," near Churchill, Manitoba, on the western side of Hudson Bay. It is a locality where the Canadian forests to the southward begin to give away to the treeless tundra that reaches northward to the Arctic.

This group in the Hall of the Birds of the World shows a scene on the upper slope of Riffelalps at timberline. In the background is a view of the Zermatt valley overshadowed by the towering snow-clad peak of Switzerland's famous Matterhorn.

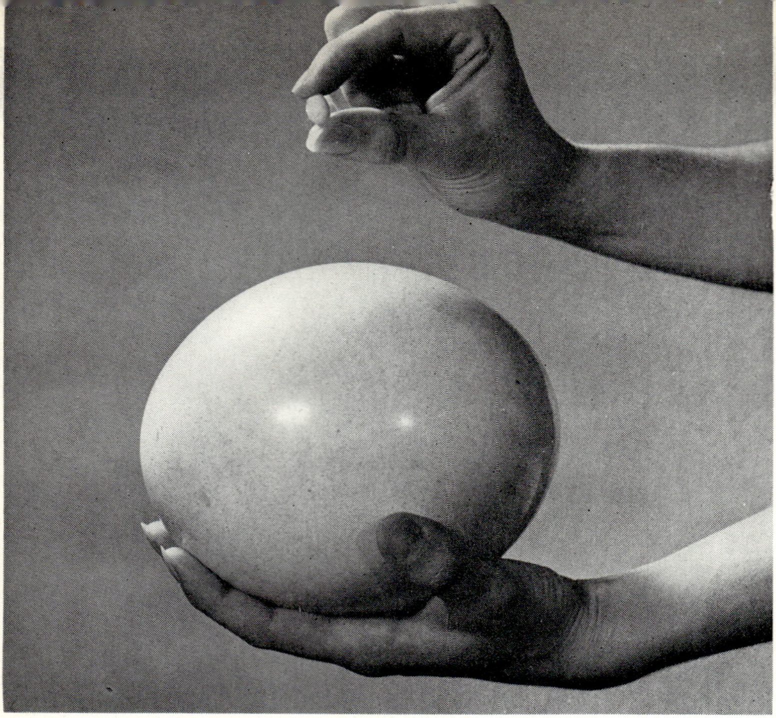

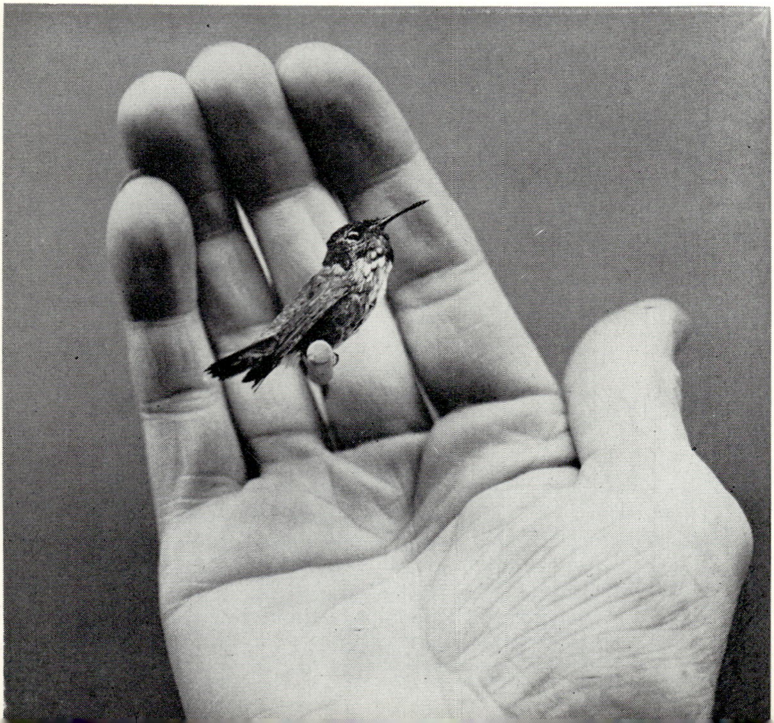

The world's largest bear is the gigantic Alaska Brown Bear, *Ursus gyas Merriam*. This pair shown against the background of mountains on the Alaska Peninsula is featured in one of the many habitat groups in the Hall of North American Mammals in The American Museum. The large male stands over eight feet tall.

Left Above: The largest and the smallest eggs of living birds belong to the ostrich and the hummingbird respectively.

Left Below: The Cuban Bee Hummingbird is the world's smallest bird and is less than two inches long.

Timber Wolves in pursuit of a deer race across the frozen waters of Gunflint Lake in northern Minnesota. This winter night drama is further enhanced by a spectacular display of the Northern Lights.

Tigers at the edge of a small pool in a grassy swamp near Karegarn on the Nepal border in India. One of the groups in the Vernay-Faunthorpe Hall of South Asiatic Mammals, this scene illustrates the manner in which the tiger's stripes blend closely with the lights and shadows of the tall grass.

Gorilla Family at Home. One of the most spectacular of the twenty-eight splendid habitat groups in the Akeley Memorial Hall of African Mammals shows the natural haunts of the Mountain Gorilla on the slopes of one of the Kivu volcanoes, Mount Mikeno, in eastern Belgian Congo. Carl Akeley himself collected and prepared the specimens. Akeley's grave lies nearby on the slopes of this mountain.

Above: "The Bridge That Walks" is demonstrated by Dr. William K. Gregory. This display utilized a Wombat skeleton, a toy steam shovel and an erector set to illustrate the principles of construction seen in the skeletons of animals. The animal's limbs correspond to the bridge towers or piers; the backbone to the bridge arch or cantilever; the head and jaws to a steam shovel; and the neck to a drawbridge.

Right: Skeleton of Horse and Man Compared. This dramatic and remarkably vital exhibit is the result of the skillful artistry of preparator S. Harmsted Chubb. So accurately articulated are the bones in these skeletons that the completed mount rivals the sculptor's art. One of the exhibits in the Hall of the Horse Under Domestication.

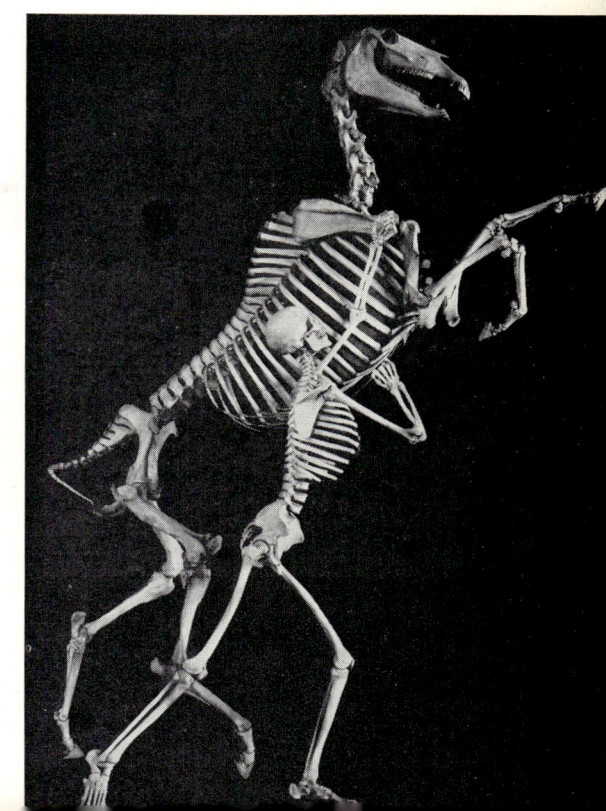

Stone Age Artists At Work. A detail of a model of the famous cave at Altamira, Spain in the Hall of Prehistoric Cultures. The complete model shows the entire cave, including its entrance high on the side of a hill. The cave walls are decorated with drawings of the animals hunted by prehistoric man. One cave artist holds a primitive stone lamp while his companion fills in the outlines of a prehistoric bison with a soft crayon made out of animal fat and brown earth.

Two Peruvian archeologists and American Museum president F. Trubee Davison (on the right) examine the skull of a mummy. The Peruvian government allowed this mummy, still wrapped in its original bundle to be sent to New York to be opened and studied by American Museum scientists. After this gesture of international cooperation among scientists, the ancient Peruvian gentleman and his burial wrappings were returned to his homeland.

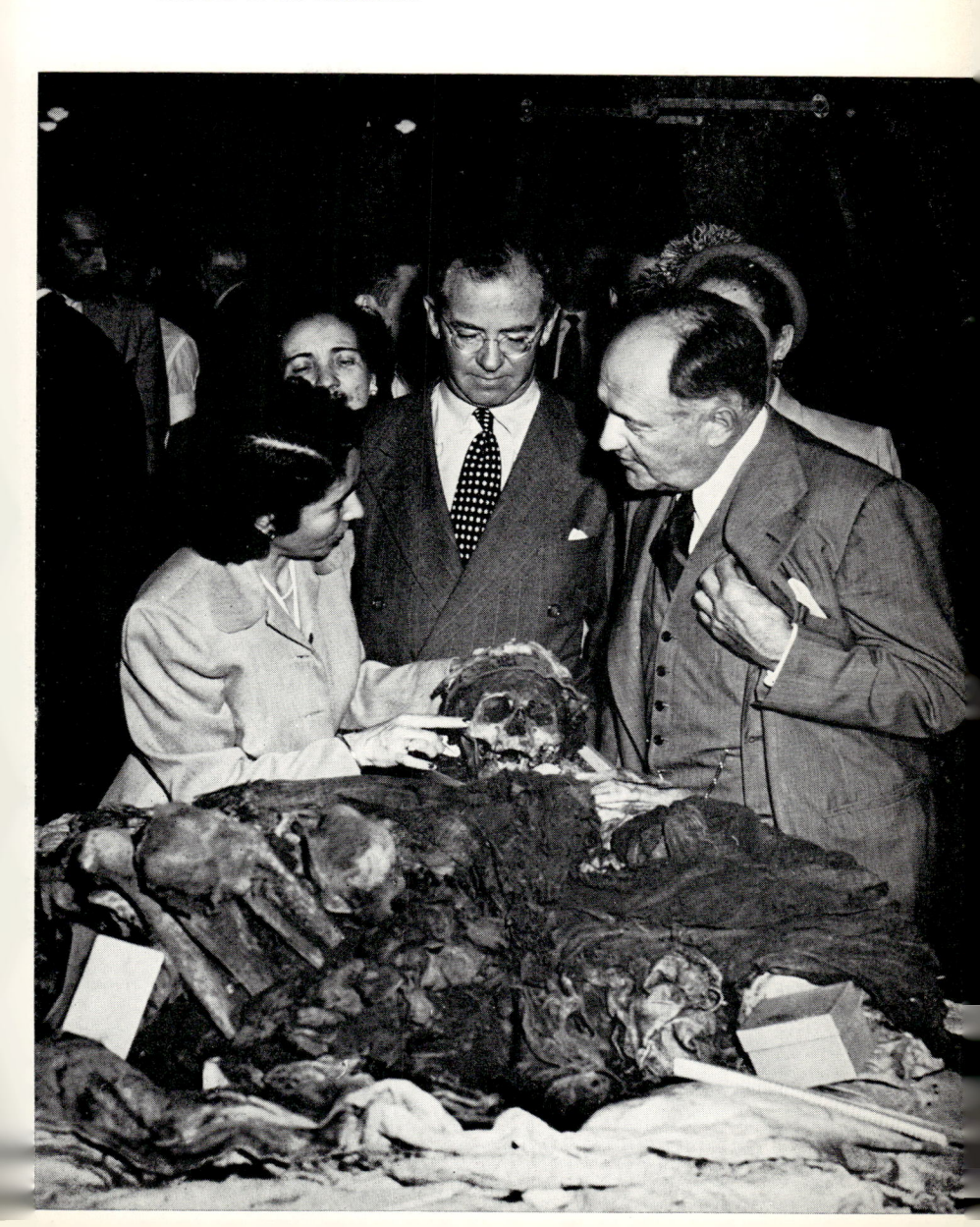

Community life in a Hopi Pueblo. One of the large groups in the Hall of the Southwest Indians shows

for corn; the larger conical roofed structure on the right is a dwelling.

A Yuchi Indian settlement in Georgia. This diorama in the Eastern Woodland Indians Hall shows a group of Indians preparing deer skins. The idea of a log cabin was introduced by the Swedes who settled

that were easy to obtain. Since birch trees were plentiful in their territory they developed many uses for the bark which they stripped from these trees. Wigwams, canoes, dishes, buckets and other utensils were fashioned out of bark. This is a miniature group in the Eastern Woodland Indians Hall.

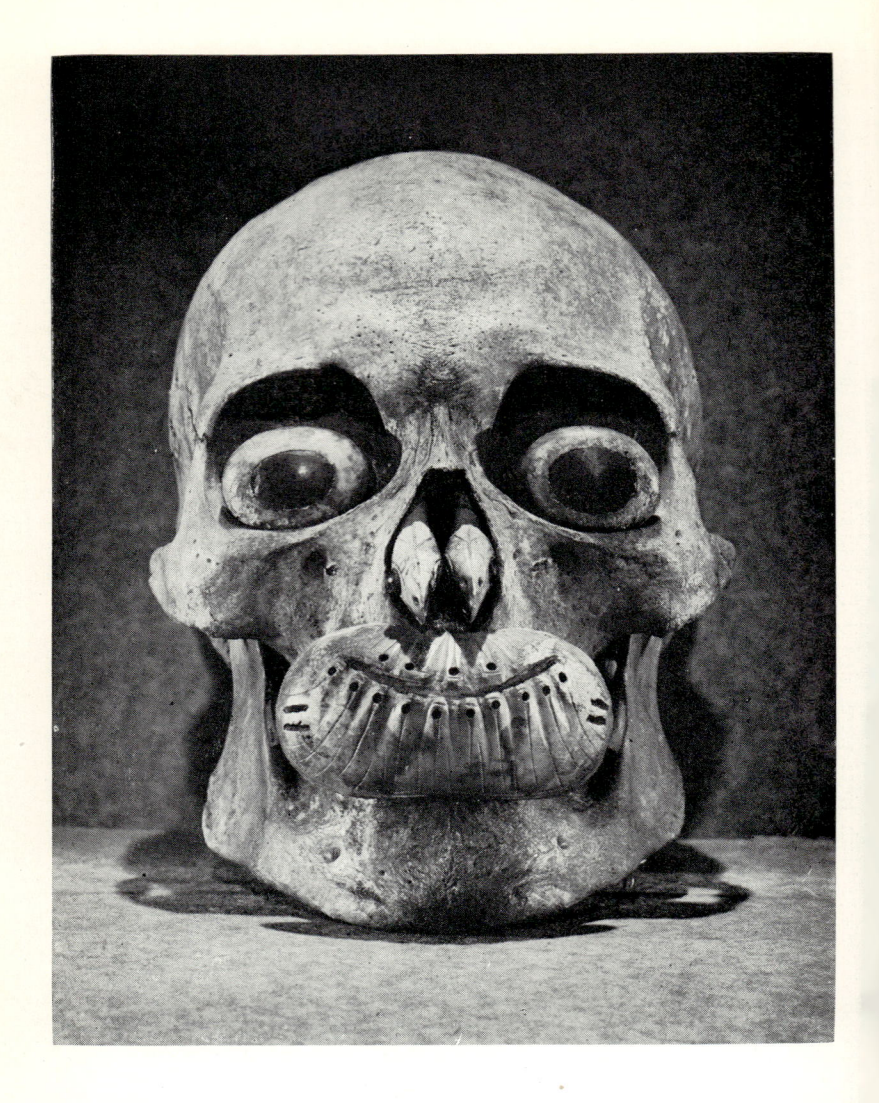

Relic of the Mystery People of Alaska. This ancient human skull was dug up at Ipiutak in northwest Alaska. The skull is fitted with ivory eyes with inlaid pupils of jet. Ivory nose plugs representing birds' heads and a carved ivory mouth cover complete the adornment of this ancient Alaskan's skull. The material culture of the Ipiutak people as revealed by the discoveries at Pt. Hope, Alaska show little resemblance to that of the modern Eskimo. These people may have been a foreign group who entered the Arctic from the south before the birth of Christ.

"False-Face" dancers from New York State. The False-Face Society of the Onandagas was and still is an important Indian community group. These two dancers wear carved wooden masks and carry noisemakers in the form of a turtle-shell rattle and a metal saucepan. The masks represent supernatural beings.

Above: A Giant Olmec style head greets visitors to the Hall of the Archaeology of Mexico and Central America.

Below: Two of the famous and mysterious Easter Island heads seen on their original site. The South Pacific Hall features a life-sized cast of one of these heads.

"Inca" Gold. These handsome examples of the skill of ancient Peruvian metalsmiths are in the Hall of the Indians of South America. Not all of the "Inca" gold was pure. Some of it was cleverly alloyed with silver. Handsome gold ornaments are often found in the mummy bundles of persons of high rank or importance.

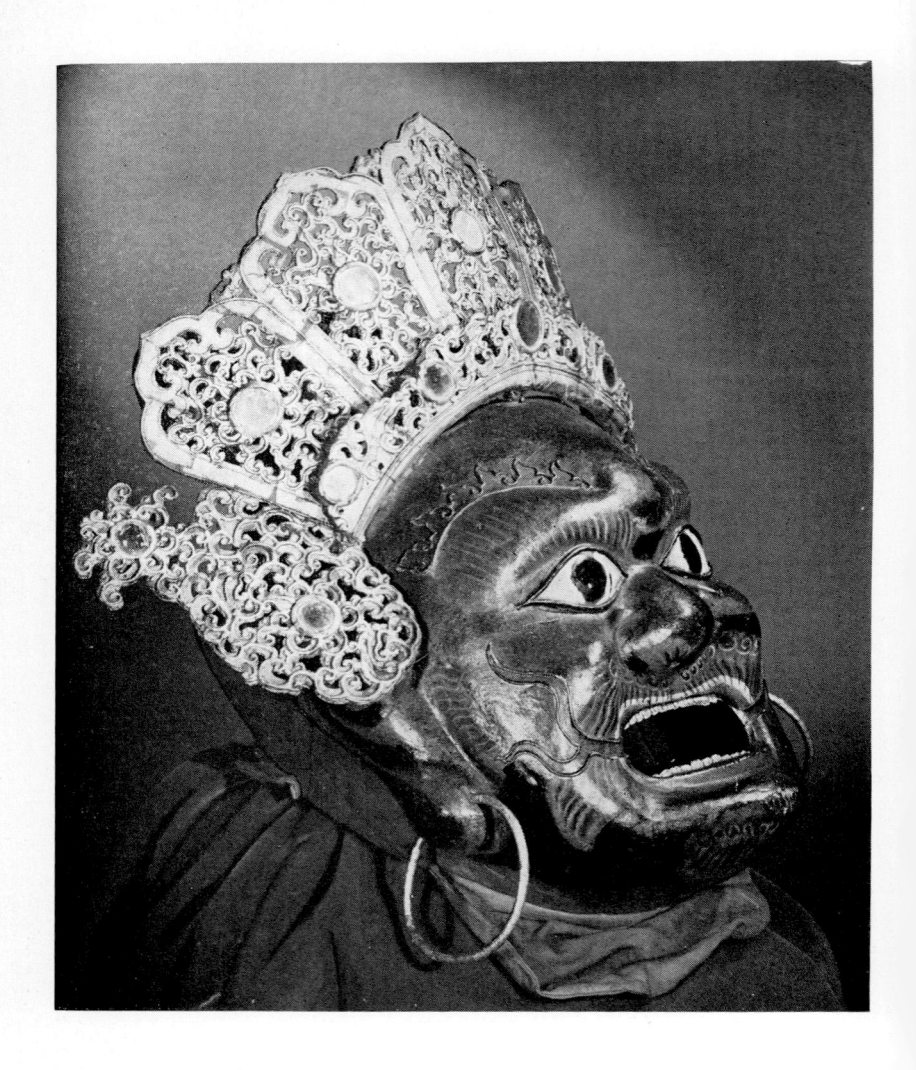

Tibetan Mask. This highly ornamented ceremonial mask is used by the Lamist priests in one of the many religious ceremonies that are so much a part of Tibetan life. The Museum's Whitney Collection from Tibet is one of the finest of its kind in this country. Most of the items in the Tibetan Hall are associated with Lamaism, the national religion of Tibet. Included in this collection are many painted scrolls, temple trumpets, prayer wheels and other religious paraphernalia.

Above: A Mangbetu Village Scene. The great slit-drum is like the one in the Museum's African Ethnology Hall.

Below: The rare and exquisitely carved ivory pieces in The American Museum's famous Lang Collection once belonged to an African king. They were carved from elephant ivory in the same manner as shown by the African craftsman in this picture.

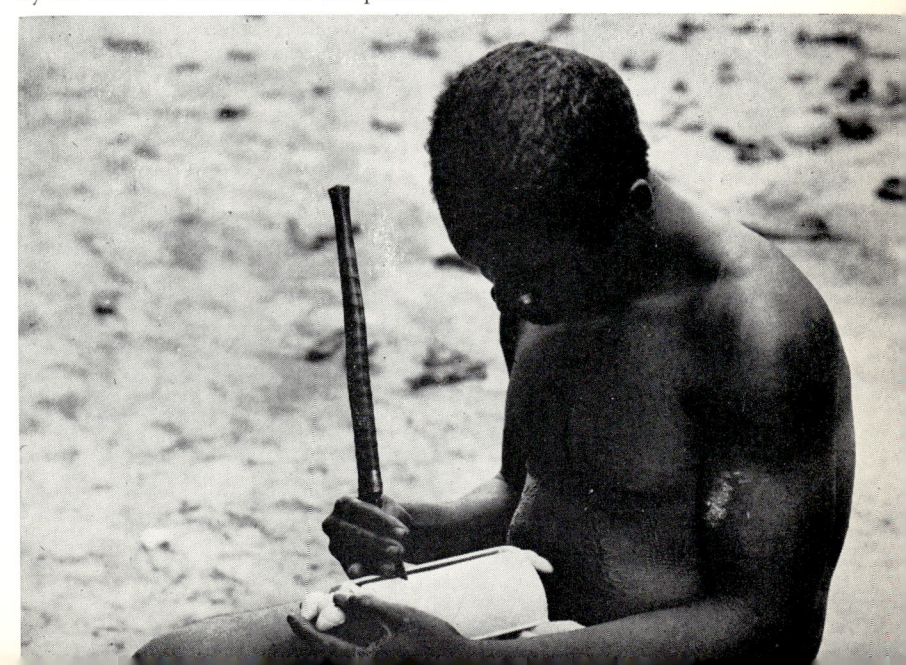

THE SKELETON FROM FISH TO MAN

The Skeleton From Fish to Man. An exhibit in the Hall of the Natural History of Man. Since the remains of most animals of the past have been destroyed by natural agencies only a few fossils have been discovered which represent creatures lying in or near the direct line of ascent from fish to man. This exhibit, prepared under the direction of Dr. William K. Gregory shows ten stages in the evolution of the vertebrate skeleton. The *First Stage* is represented by *Cheirolepis,* a fossil fish from the Devonian Age. *Stage Two,* another fossil fish named *Eusthenopteron* from the Upper Devonian represents a long step forward. This prehistoric fish had two pairs of paddles corresponding to limbs. The *Third Stage* is typified by a fossil reptile, *Diplovertebron,* from the Carboniferous Age. *Diplovertebron* is the oldest known type of four-footed animal. It was also the first known creature to have five digits on each of its fore and hind limbs.

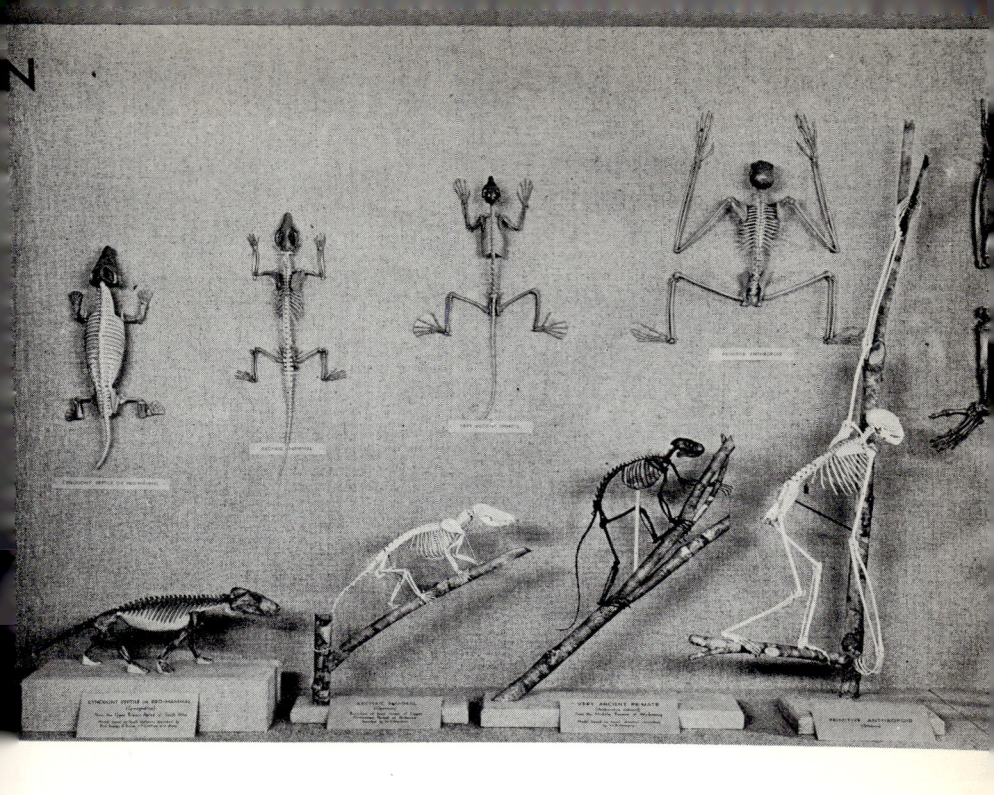

A primitive lizard-like reptile, *Seymouria,* from the Lower Permian Age represents the *Fourth Stage* in the evolution of the skeleton. *Seymouria's* limbs were better developed than those of the *Diplovertebron,* yet it still held its body close to the ground as its legs sprawled outward. *Stage Five, Cynognathus,* was a mammal-like reptile from the Upper Triassic of South Africa. Its limbs were better adapted for running as they were placed more directly under the body which was thus held higher above the ground. The structure of the skull, backbone and limbs of *Cynognathus* has many features which approach those of the mammals. The *Sixth Stage* is represented by the skeleton of the opossum, a modern mammal. The opossum, however, still retains most of the important features of older fossil mammals. It has a low type of skull, teeth and brain but its grasping hands and feet enable it to climb trees.

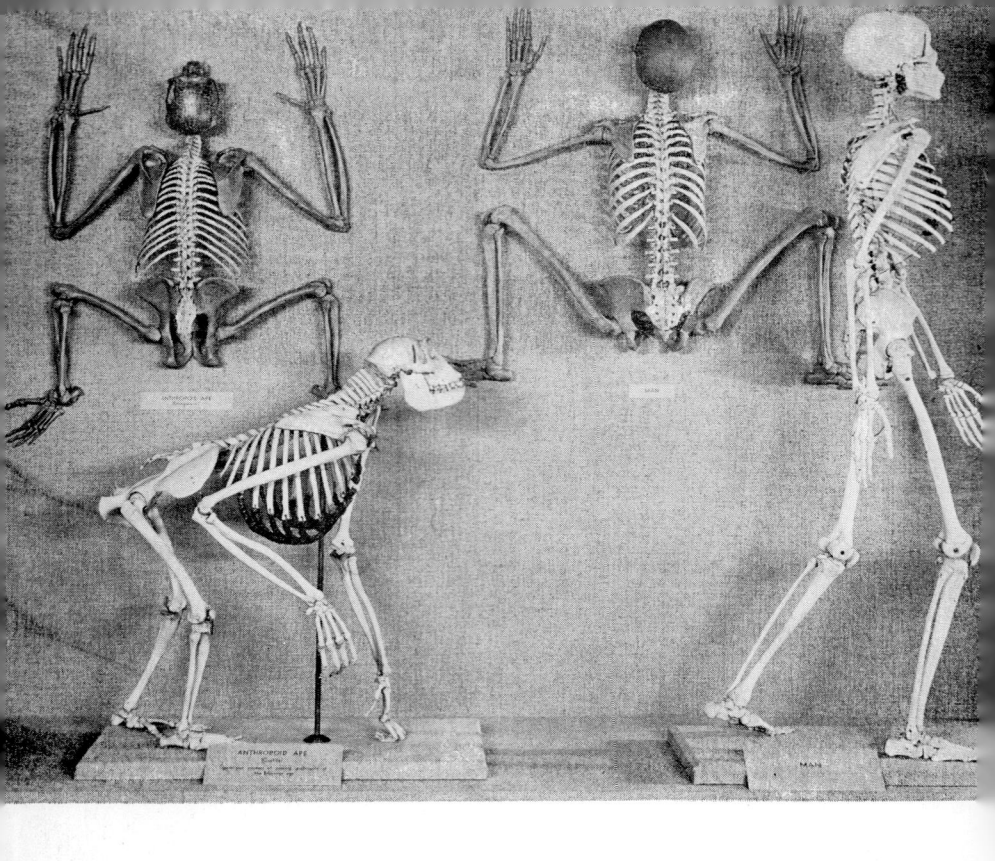

Notharctus, a small primitive primate typifies *Stage Seven.* This little prehistoric creature, lying near the lower limits of the order of Primates is well adapted to a life in the trees. *Notharctus* had larger eyes and a larger brain than any of the earlier primates. The last three stages in the evolution of the skeleton are represented by present-day primates. *Stage Eight,* the gibbon, is a tree-living descendent of the first of the man-like apes. The gibbon walks on its hind legs when on the ground. The gorilla and the chimpanzee represent *Stage Nine.* The brain of these apes is much better developed than any of the other creatures excepting man. Man represents the final or *Tenth Stage.* The human skeleton is constructed upon the same overall plan as is that of the gorilla, the chimpanzee and the gibbon. In Man, however, the backbone, pelvis and limbs are modified to allow him to walk erect. This frees Man's arms for other uses. The human brain is much larger and is better developed than the brain of any other primate.

In the beginning most of the specimens of birds the Museum accumulated were by similar gifts and purchases. In 1932, through the great munificence of the Whitney family, the Museum acquired what was then reputed to be the most studied and therefore the most useful single collection of birds in the world, the famous Tring Collection owned by Lord Rothschild and housed in a special museum he had built for it at Tring, England, containing about one-quarter of a million birds and constituting about one-third of the Museum's present collection. The Museum is further indebted to the generosity of the late Harry Payne Whitney for the entire wing of the Museum which houses the offices, laboratories and new exhibition halls of the bird department.

Today the bird exhibits fill four large exhibition halls and in addition there is a special gallery for bird paintings and a corridor of Auduboniania, as well as a much used corridor exhibit of local birds and an exhibit of eggs and nests. Not in the main bird halls but nearby is the Bird Sanctuary Group in the Theodore Roosevelt Memorial Hall. The vast study collection is housed in the Whitney building in hundreds of specially built easily accessible storage cases. Convenient work tables and benches are scattered throughout the rooms in which the study collections are housed for the convenience of the staff and visiting ornithologists who must constantly use its valuable specimens.

The best introduction to birds the Museum exhibits provide is to be found in the recently opened Sanford Hall of the Biology of Birds. This hall, designed to tell the story of birds and of their relation to man, is the gift of Dr. Leonard Sanford, long a good friend of the Bird Department and the Museum. The late Dr. Chapman explained the function of the Hall in the introduction to a booklet he prepared describing the exhibit. "There are many parts of the earth occupied by birds where man is unknown," he wrote, "but there is no place inhabited by man where birds are not also found. Their flesh, their feathers, their forms and flight, their habits, their food, their travels, their songs and companionship all may enter into our lives. From the beginning of this association birds have been essentially as they are today but man has risen from the primitive condition, in which some races still remain, to the high stage of civilization his leaders have now reached. We believe that a review of our present and past relations with birds will show that they have played, and continue to play, an inestimable part in both our mental and physical existence. This review we have attempted to make in this exhibit."

The Sanford Hall, still not completed, goes far beyond merely pointing out the relations between birds and man. It contains a comprehensive display of the principal genera of birds and gives an indication of the bird's position in vertebrate evolution. Special sections deal with rare and extinct birds while other alcoves treat such aspects of ornithology as migration, range, distribution, reproduction and evolution. Separate groups show some of the larger examples of nests of solitary and social birds. Center floor cases contain detailed exhibits on feathers, color, flight and bird physiology and there is one case devoted to fossils.

A great habitat group showing some of the large tropical birds in flight against a vivid sunset catches the visitor's eye at the entrance to the hall. Here are great herons, ibis, spoonbills and other varieties of some of the spectacular long-legged waders of the American tropics. The visitor can then go to his right and commence the review of bird life by studying the exhibit called "Birds and Man" and thus discover the importance of these feathered creatures in his life, or he can go to the left and see how birds evolved from their reptilian ancestors and proceed to see the important representatives of the feathered world.

The "Birds and Man" alcove has been organized to present in ten sections, ten ways in which birds and man are associated. Beginning with "Birds as Food," and continuing with "Birds as Clothing and Ornament," "Birds as Omens and Symbols in Mythology and Tradition," "Birds as Emblems and Insignia," "Birds in Books," "Birds in Art," "Economic Importance of the Feeding Habits of Birds," "The Domestication of Birds," "Commercial Uses of Birds," the series ends with a section on the "Sentiment of Birds" which reminds the viewer of the beauty of bird song and the pleasures of bird watching.

A number of species of birds have become extinct since the time when their prehistoric ancestor, the *Archaeopterix* flew about in the Cretaceous forests millions of years ago. Some of these extinctions came about as the normal process of evolution and change, but some have been the direct result of man's avariciousness and stupidity. One of the prime examples of the latter is the instance of the passenger pigeon, great flocks of which, numbering billions, once darkened the North American skies. They lived mostly in the eastern United States and there are still many Americans living who remember them. There were so many that no one believed that they could ever become extinct. Alexander Wilson, the "father" of American ornithology tells of visiting in Kentucky in 1806 one of their breeding places which was several miles wide and more than forty miles long. In this forest tract the trees were laden with passenger pigeon nests and it was not uncommon for one hundred nests to be found on a single tree. Wilson also describes a flight of pigeons which he saw. There were so many birds that the sky was darkened by their wings in every direction. The great cloud took four hours to pass overhead and Wilson estimated that the flock must have been stretched out some two hundred and forty miles in length and that it was a mile wide. He also estimated that it must have contained over two billion birds.

Today, not a single living passenger pigeon exists. Only in museums and in private collections can one find mounted specimens of the passenger pigeons and these are becoming scarce. We may not have conclusive evidence to prove that man was entirely responsible for the disappearance of these creatures but we can say with certainty that he played the greatest part in their extinction.

The Indians used to hunt passenger pigeons, just as they hunted bison, taking just enough to meet their needs. The great slaughter began when a market demand for the birds was created by the appearance of the white men. Armed with guns, clubs, stones and nets, the great flocks were set upon whenever they came near human habitation and the birds were slaughtered and their bodies sold as food. Audubon says that in 1805 boatloads of these birds were sold in New York City at the price of a penny a bird. Millions and millions of birds were killed year after year. What was particularly bad was that they were slaughtered at all times even during the mating, roosting and nesting seasons. The end had to come some time. In 1879 the last large hunt was held in Michigan and in this raid no fewer than a billion birds were killed. The flocks began to diminish noticeably, laws were finally passed restricting their killing but the laws were not enforced. The destruction of the great forests where the birds lived hastened their end and finally no more large flocks were seen. The relatively few survivors roamed the countryside, fair game for shotguns, and then the inevitable day dawned. After 1908 there were no more wild passenger pigeons to be seen anywhere. In 1914 the last passenger pigeon known to exist died in the Cincinnati zoo.

The Passenger Pigeon Group in the Sanford Hall shows a small flock of these rare birds roosting on an oak tree. Nearby are cases containing other birds that have vanished since the coming of man, such as the great auk, the dodo, the Labrador duck and the heath hen. The American Museum of Natural History's ornithologists have always been alert to the threats to rare species of still living birds and have done their part to inform the people and to press for action that will insure us against losing any more of our avifauna.

On the second floor of the Whitney wing is the Whitney Memorial Hall of Pacific Bird Life, containing eighteen large habitat groups of Pacific birds. The exhibits will ultimately represent the Pacific Ocean by means of specimens from selected localities which extend from Laysan Island north of Hawaii south to the sea off New Zealand, on the verge of the Antarctic and from the Galapagos Islands in the east to the Philippines, New Guinea and the Australian Barrier Reef in the west. The exhibits in the hall are arranged to correspond with the actual compass directions. At the south end, where the visitor enters, are the Southern Pacific groups, and as the visitor works his way to the north end of the room he passes a series of groups each representing a more northerly latitude. On the east side of the hall are arranged specimens chiefly from the easterly half of the Pacific; those on the western side of the hall are from the western Pacific regions. Overhead, suspended from the dome ceiling which has been painted to represent the sky, a similar succession is presented by such birds in flight ranging from the Antarctic snow petrels and whale-birds to equatorial man-o'-war birds and tropic birds.

The location of each exhibit has been indicated on large decorative mural maps and charts in the entrance corridors, which also contain information concerning ocean depths, prevailing winds and currents and other geographic matters. One of the maps shows the courses of some of the vessels that have become notable in the history of Pacific science. Here are shown the voyages of the pioneering ships "Resolution" and "Discovery" commanded by Captain James Cook (1776-80); those of the Russian "Vostok" and "Mirnyi" (1820-21) under Bellinghausen; the famous cruises of the "Beagle" (1832-36) and the "Challenger" (1874-75); the surveys of the United States Fish Commission steamer "Albatross" (1888-1910) and finally the extended wanderings of the American Museum schooner "France" (1922-32), the floating home and headquarters of the Museum ornithologists when they collected birds for the study collections and for Whitney Hall. In addition to the charts there is a special memorial niche at the entrance which bears an inscription of dedication and which contains a bust of Harry Payne Whitney by the sculptor Jo Davidson and another of Mr. Whitney's father, William C. Whitney, by Augustus St. Gaudens.

SCALES, FEATHERS AND FUR

The Ship-Followers Group presents a realistic scene of the open ocean south and east of New Zealand as seen from the rolling and pitching deck of an old-fashioned sailing vessel. Painted on the background can be seen the Whitney South Sea Expedition schooner, the "France." This part of the South Pacific Ocean is known as the "Roaring Forties." A day in mid-summer scene (February) is represented, and some of the sea birds that follow ships for food thrown overboard or stirred up by the ship's wake are shown. The birds include the sooty and wandering albatrosses, Cape pigeons, whale-birds, Mother Carey's chickens (stormy petrels) and other petrels.

The Peruvian Guano Group, a scene at the famous Chincha Islands at the Bay of Pisco reminds the visitor of the days gone by when the valuable bird droppings which had accumulated on these islands over hundreds of years proved a magnet for the guano hunters who shipped it away to be sold as fertilizer. The guanay or guano-producing cormorant, the boobies and pelicans are the three most important birds depicted in the group. It is this trio of sea-birds which produce annually in this rainless region more than one hundred thousand tons of marketable fertilizer. This guano fertilizer has been estimated as being worth 33 times its weight in farmyard manure. The guano business in Peru is the largest industry in the world that is based upon the conservation of wild birds. The Peruvian Government realizes the value of these birds and takes good care to see that nothing happens to destroy them and the industry they constitute.

Other groups in the Whitney Hall show bird life in the Galapagos; the Island of Kauai in the Hawaiian Islands;

Tuamotu, the coral atoll; the Marquesas; showing the valley of "Typee" famous as the locale of Herman Melville's romance of the same name; the Snow Mountains of New Guinea; Fiji; Little Diomede Island in the Arctic; and other typical localities.

The Solomon Island Group has a dramatic and memorable importance to the visitor since the United States armed forces made history. This Museum group shows a peaceful scene featuring the bird life of a hot, humid, and mountainous archipelago. In the foreground is a cluster of native huts, and a garden in which coconut palms, bananas, papaya, cassava, breadfruit, taro, and sweet potato are growing on the site of a tropical forest recently cut down by the natives. Doves, parrots, lories, and cockatoos are seen along with a megapode or brush fowl, which lays its eggs in mounds of rotting vegetation where the heat of fermentation hatches them. The background shows Guadalcanal viewed across a narrow inlet. Flying over Guadalcanal is a handsome specimen of the Brahminy kite, a bird of prey.

This group was opened to the public at just about the time word came to the States of the successful action on Guadalcanal. When this and the other groups showing scenes in the South Pacific were put on display many folks who had sons in action in that theatre of operations came to the Museum especially to "see where their boys were." When one such couple came and were taken to see the Solomons group and were told that the bird in the air over the Island of Guadalcanal was a kite, the father turned to his host and guide and said with proud tone and a twinkle in his eye, "You call that bird a kite? It looks more and more like an American eagle the longer I stand here."

The New Guinea Group shows a scene on Lake Habbema, 11,000 feet above sea level with Mt. Wilhelmina, in the Snow Mountains, as a background for a number of Macgregor's birds of paradise with the orange-colored shields behind their eyes, and the beautiful, iridescent-breasted splendid birds of paradise. More of these exotic creatures are installed in cases in the center of the Whitney Memorial Hall. These birds are found only in the New Guinea area. Centuries before their real habitat was known to western ornithologists, skins had made their way into the Orient via traders and seafarers, and into Europe when El Cano, who took charge of the first circumnavigation of the world after Magellan had died, brought back to Spain some which he had received from the Rajah of Batjan in the Moluccas. They had been treated by the New Guinea natives who took out the body of the bird, cut off its legs and dried the skin over a fire.

This confused the Europeans and some amusing accounts were written concerning the birds. In 1551, Cardano wrote, "The 'Birds of the Gods' come from the Moluccas where they are picked up dead on land or in the sea. They are never seen alive. Since they lack feet, they are obliged to fly continuously and live therefore in the highest sky far above the range of human vision . . . the back of the male is hollowed out and the female lays her eggs into this cavity and incubates them while the male continues flying . . . they require no other food or drink than dew from Heaven."

The Bird of Paradise was so called because it was first believed that Paradise was its true home. What we know about these birds today certainly is not as romantic a story as that of Cardano's. Except for their remarkable and exquisite plumage, they are like all other birds, they have legs, they lay their eggs in nests, their food is far more prosaic than "dew from Heaven" and they are not exceptional distance flyers.

Starting about 1880 and building up to a peak in 1910, the hunting of these birds for their plumage for mi'lady's hat became a serious threat to their existence. Before the first World War nearly 10,000 skins were exported annually from German New Guinea alone. At last conservationists became alarmed at the situation and finally, in 1924 their destruction was forbidden by law. Since that time they have been hunted only illegally.

Twenty-five years ago the American Museum's collection of birds of paradise was a meager one. In 1928 the generosity of Museum Trustee Dr. Leonard C. Sanford made it possible to send Rollo Beck to northeast New Guinea and Dr. Ernst Mayr to Dutch New Guinea to obtain specimens. In 1929 these ornithologists were followed by the Whitney South Sea Expedition which concentrated on British Papua. As a result the Museum assembled the finest bird of paradise collection in America but it was still a very poor second to the collection at Tring which Lord Rothschild had spent 40 years building and which contained nearly every known species including seventeen which he described himself. When, in 1932, the Rothschild Collection was put up for sale and was purchased for the Museum by Mrs. Harry Payne Whitney, the Museum's collection of these wonderbirds became the finest in the world.

It is an ironic bit of natural history that these resplendent creatures appear to be close relatives of the crow family. The most primitive look very much like small crows and like the crow, most of the birds of paradise, despite their unexcelled beauty have raucous and unmelodic voices.

The Birds of the World Hall on the second floor was designed to show by means of twelve habitat groups, the major faunal areas of the world and their characteristic birds. Eleven of these groups, which have been completed, show the pampas and lagoons of the south temperate zone of South America; the Paramo zone of South America featuring Mt. Acancagua, the highest peak in the Americas; Barro Colorado Island in the Canal Zone where for many years a biological field station has been maintained by the Institute for Research in Tropical America; the island of South Georgia in the sub-antarctic zone; the grassy country of the East African plains exemplified by the Kidong Valley, near Nairobi; the equatorial forest along the Congo River in western Africa; the eucalyptus-dotted savanna in the Blue Mountains of New South Wales; the shores of Tsagan Nor, the white lake in the Gobi Desert near the Altai Range; the Zermatt Valley and the Matterhorn in Switzerland representing the Alpine zone; the New Forest in Hampshire as typical of the Old World north temperate zone; and finally the Tundra Group which shows a scene on the western side of Hudson Bay, near Churchill, Manitoba.

The New Forest Group is of particular interest to American bird enthusiasts as here they can see some of the birds immortalized in song and story by English writers. The skylark is shown on the wing, the magpie and dozens of other birds of the English countryside are seen bedeviling an owl which they have discovered perched in an ancient beech tree. The birds seem to be fearless as they torment the owl, but when night comes, the owl once more becomes a dreaded foe.

Many of the species are the same as those in the North American north temperate zone but the local names of the birds differ. In some cases the same name is applied on both sides of the Atlantic to quite different species, as in the case of the European robin and the American robin.

This group has another significance. When Theodore Roosevelt was President of the United States he still indulged his keen interest in bird study. While entertaining Lord Grey of Falloden, Roosevelt took him on a bird walk, and Grey, also a bird enthusiast, offered to return the compliment should Roosevelt ever visit England. In 1910, Roosevelt visited Europe and made a triumphal tour. Reaching England with his footsteps dogged by reporters and crowds of admirers, Roosevelt suddenly disappeared for a few days. No one knew where he had gone until later it was revealed that he had thrown the reporters off his trail and had allowed his friend, Lord Grey to take him on a bird walk in the New Forest.

The Museum Group shows a part of the New Forest in the Valley of Itchen, in Hampshire, where these two men watched the birds together. The label of the group in the Museum reproduces part of the page of the register of the small inn where these two amateur ornithologists spent the night. Lord Grey of Falloden was once asked what he thought of the American Museum of Natural History. He remarked that he would gladly cross the Atlantic just for the sake of being able to spend a few hours in its halls. The group is dedicated to Lord Grey and was the gift of Mrs. Carl Tucker.

The first hall of the Museum to be entirely devoted to habitat groups was the Hall of North American Bird Groups on the third floor. This hall and the Hall of Birds of the World are the only two large bird halls which are not located in the Whitney Wing. There are thirty habitat groups in the North American Bird Hall showing bird life in natural settings from all parts of North America, from the Bahamas and the southern end of the Mexican Plateau to Hudson Bay and Bering Sea, and from the Atlantic to the Pacific. Some 90,000 miles were traveled to secure specimens and material for the groups. The excellent background paintings were by such artists as Bruce Horsfall, Hobart Nichols, Charles Hittell, Louis Agassiz Fuertes, and Carl Rungius.

The groups were made possible through the generosity of a number of friends of the Museum including J. L. Cadwalader, Mrs. Morris K. Jesup, Mrs. Philip Schuyler, J. B. Trevor, Mrs. Robert Winthrop, and Messrs F. A. Schermerhorn, H. B. Hollins, H. C. Pierce, H. W. Poor, C. Brandreth and J. C. Carter.

The hall was started in 1898 and for over a half century these pioneer habitat groups of birds have continued to delight and instruct the Museum's visitors. It is a tribute to their designers and builders that the groups have withstood the ravages of time so well. The groups in this hall present a panorama of North American bird life which includes scenes from Vera Cruz, Mexico; Cobb's Island, Virginia; the Palisades of the Hudson; the Jersey meadows; Florida; the coast of South Carolina; the Potomac River near Washington; Tucson, Arizona; southern California; California; Andros Island; Bahamas; Southampton Island, Hudson Bay; Heron Lake, Minnesota; Bates Hole, Wyoming; Ptarmigan Lake, British Columbia; Medicine Bow, Wyoming; Halsey, Nebraska; Crane Lake, Saskatchewan; Lake Umbagog, New

Hampshire; and Bird Rock in the Gulf of St. Lawrence. Most of the specimens were collected by the late Dr. Frank M. Chapman, who was in charge of the Museum's Bird Department for many years and did much to popularize bird study in America. His "Handbook of North American Birds" was the "bible" for several generations of American bird watchers.

Not a dramatic exhibit in any sense of the word, one of the most thoroughly used exhibits of birds at the Museum is that on "Birds Within a Fifty Mile Radius of New York." This series of corridor cases contains mounted specimens of practically all the birds that can be found in the New York area. Special labels, keyed in color, indicate the migrants, the summer and the winter visitors and the permanent residents. At almost any time of day, especially during the spring and fall migration seasons, one can see bird enthusiasts standing before these cases peering intently at their contents, sometimes checking now and then with a bird guide or with their own notes, evidencing how many people from all walks of life find a refreshing and stimulating hobby in bird watching. Many are business and professional people who go to Central Park during their lunch hours and even early in the morning, before reporting to their offices, to search out and identify and add to their records the small feathered migrants. When they are not certain of the identification of a bird they have just seen, they will often cut across the park to the conveniently located Museum and verify their identifications by checking on the mounted bird in the case. Though the Museum is proud indeed of its beautiful Whitney Memorial Hall with its remarkable habitat groups and its glamorous birds of paradise, and though it may boast of the excellence of the Sanford Hall of the Biology of Birds, the small Local Bird Collection also does its part in fulfilling an important function of the Museum, namely that of being "an instructive and acceptable resort for the people of our city."

One day the late Dr. Allan Dafoe, the modest little Canadian country doctor who had helped bring the Dionne quintuplets into the world, came to visit the Museum. Dr. Roy Chapman Andrews, then Director, took personal charge, and saw to it that Dr. Dafoe enjoyed his visit. As they walked about the exhibition halls, Dr. Andrews would frequently question his guest about the five babies he had delivered. They came to the Hall of Ocean Life and standing at the balcony edge, Dr. Andrews pointed down to a specimen which was hanging from the ceiling, and turned to Dr. Dafoe and said, "You, Doctor, have told me all about your prize babies, now let me show you the Museum's prize baby." The baby which the two men then examined was a model of but one of the many thousands of specimens of present-day mammals that come under the care of the Museum's Department of Mammals. It was a baby sperm whale, about two weeks old, which was the only whale ever to be brought into the Museum in the flesh. This young whale had apparently lost its mother and had wandered into New York Harbor, slipped past the busy boat life of the bay, swam up the East River and finally came to grief in the Gowanus Canal of Brooklyn. Some fisherman there succeeded in capturing and killing the unfortunate creature and the Museum's mammalogists were quick to purchase the carcass and cart it back to the Museum.

Whales are big creatures, but they are by no means the

largest of the responsibilities of the Mammal Department which has the task of maintaining the bulky study collections and the many exhibitions of the world's array of its dominant form of animal life, the mammals, which range in size from whales a hundred feet in length and weighing over 130 tons, to tiny mice and shrews a scant few inches long and a mere couple of ounces in weight. To a mammalogist, a mouse can be no less important than a whale. Each represents to him special problems and special meanings. The average visitor is more apt to be impressed by great size and strength. The Mammal Department's exhibits contain a sufficient assortment of the larger and more prepossessing. representatives of this important group of animals.

Like its early bird collections, the Museum's mammal collections were among the first to be acquired and to be placed on exhibit in the old Arsenal building in Central Park. The deep roots of this department go back to 1869 when Albert Smith Bickmore was the first superintendent of the Museum. Bickmore was then in charge of mammals, birds and other branches of zoology with Dr. Holder as his assistant. Professor Whitfield headed the only other division the Museum then had, that of Geology.

In 1872, Holder devoted his attention to the zoological collections and was made Curator of Zoology in 1881. In 1885, the Zoology Department was divided into two Departments, one of Mammals and Birds, the other of Marine Zoology. Holder then undertook the Curatorship of the Department of Marine Zoology. He had been an associate of Professor Agassiz and had conducted an important investigation of the fauna of New England waters and later on the fauna of the Florida shores.

SCALES, FEATHERS AND FUR

At the same time that Holder was made **Curator of Marine Zoology**, Dr. J. A. Allen was made Curator of the Department of Mammals and Birds. Allen, like Bickmore, had studied under Agassiz and had served as an assistant in ornithology in the Museum of Comparative Anatomy at Harvard. It was through his efforts that the Department of Taxidermy was started in 1886. Allen also started the first study collections in his field of ornithology at the Museum and it was from his Bird and Mammal Department that the first expedition organized for collecting mammals and birds was sent out. He made numerous contributions to systematic zoology and was recognized as one of the foremost authorities on mammals and birds in his day.

Other men who joined this combined department of mammals and birds were Frank M. Chapman (1888), W. De Witt Miller (1903) and Roy Chapman Andrews (1907). Chapman later became Curator of Birds when the Department was split in 1920 into two Departments, the Bird Department and the Mammal Department. Miller, whose untimely death in the prime of his life robbed the American Museum of one of its more promising staff members was an ornithologist while Roy Andrews, later to make his mark as the explorer of the Gobi Desert and eventually to become the Museum's Director, was an assistant in mammalogy.

When Allen assumed charge of the Museum's mammal collections the specimens numbered about 1,000 mounted skins and 300 mounted skeletons. All of these were on exhibition and there was no study collection. Today the Mammal Department has four large exhibition halls and also the bulk of the exhibition space in the Hall of Ocean Life filled with its exhibitions of marine mammals. The Museum's present collection of mammals, containing almost 150,000 specimens, is second largest in the New World, being surpassed in size only by that in the United States National Museum. But though it is not the largest, the American Museum's collection is a broader one than that of the U.S. National Museum's, the latter being chiefly North American, while the American Mammals but boasts the best collections anywhere from Central America, particularly Costa Rica and Guatemala, and the largest collection from South America in the United States, ranking with that of the British Museum's. Its representation of African mammals is probably the best in the country and its collection from Asia is one of the two top such collections in the United States.

The Museum collection of European mammals does not rank so high as do those from other areas. The best such collections are, as one might logically expect, in European Museums. But because of its expeditions to the South Pacific islands, to Australia and to New Guinea, the Museum possesses outstanding collections of mammals from these areas. Because most mammals are fairly large creatures and because there are not so many different kinds of them as there are of other forms of animal life, most mammalian species have already been described by man. New species are discovered from time to time, usually the smaller ones living in places not frequently visited by white men. However, these discoveries do not compare at all in number or frequency with the finds of "new" insects or of other more prolifically specied groups.

A large museum has its complications; it has many problems which institutions that are small or limited in scope do not even begin to experience. The very reasons for some of these complications also explain certain advantages large institutions have over their smaller counterparts. Limitations of space and funds, for instance, generally make the accumulation of a collection of cetaceans impossible for small museums.

The tremendous size of the American Museum, and the scope of its program, have enabled it to amass a considerable collection of whales and porpoises. There are no "stuffed" whales in the Museum. Clever as have been its taxidermists, the problem of preparing and preserving the skin of one of these giants of the sea has yet to be solved. Thus, the bulk of the Museum's study collections and exhibition specimens of cetaceans consists of skeletal material. Of course the records of the Department contain valuable and pertinent information regarding most of the specimens, including gross measurements and real and estimated weights ascertained at the time when the specimens were caught.

The Hall of Ocean Life houses the bulk of the cetacean exhibitions as well as the habitat groups of sea mammals other than whales. The showpiece of the cetacean collection, at least as far as many visitors are concerned, is the model of a sulphur-bottom or blue whale which hangs suspended from the ceiling of the Hall of the Biology of Mammals on the third floor. The whale from which this life-sized 76 foot model was copied was captured in the waters off Newfoundland and the museum mammalogists and preparators made careful measurements and color-notes so that they could reproduce it in wire, plaster and paint. This whale is the largest animal in the Museum and represents the largest kind of animal which has ever existed.

At some time or other in their life history **all mammals are** covered with hair, although in some cases, such as the whale and man, this coat of hair is lost before birth. Like the birds, a mammal has a four-chambered heart, which is an advantage over animals with more primitive hearts as it keeps the well-oxygenated blood from the lungs completely separated from the more poorly oxidized blood which enters the heart from the body circulation. The greatest advance over all the other forms of animal life is in mammalian brain development, with its large cerebrum, or fore-brain, proportionately larger than that of any other kind of animal. This increase in the size of the "thinking" part of the brain accounts for the greater intelligence of all mammals, including man.

The Biology of Mammals Hall with its separate alcoves each dealing with an order of living mammals, its numerous physiological models, and its displays of comparative anatomy serves as a good introduction to a study of this highest class of the phylum Chordata. Dwarfed by the model of the giant blue whale, the skeleton of Jumbo still towers above any of the other mammals in this hall with the exception of the whale. Jumbo, the African elephant and his exhibit mate Samson, the Indian elephant, both were feature attractions in the circuses of that peer of American showmen, the incomparable Phineas T. Barnum. Their skeletons still serve, if not to amuse, at least to instruct the millions of people who come to see the Museum's "show." Jumbo remains the most famous elephant of all time. So great was his fame that his name has become a household word, a synonym for the word gigantic.

Other zoo and circus animals have been donated to the Museum, among them "Samson," the elephant, "John Daniels," the gorilla, "Tip," the Asiatic elephant, and "Hannibal," the lion. In its early years the Museum was happy to get such specimens, but actually animals which have been bred or kept in captivity undergo physical changes because of the change in diet and other environmental factors. They are different from their wild brothers and the Museum has tried to exhibit Nature as it is found unaffected by man and his ways. Thus, the creatures whose mounted skins are seen in the general exhibits and in the habitat groups are nearly always specimens that were caught in the wild. However, before it was active enough to maintain collecting expeditions and while its finances were more limited the Museum was grateful to receive and to place on exhibition specimens from zoos and circuses. Such specimens are still brought into the Museum but their skins and skeletal material are used for study purposes and the animals do not appear as exhibition specimens.

In addition to the Hall of the Biology of Mammals the Museum mammal exhibition halls include the well-known Akeley Memorial Hall of African Mammals, the North American Mammal Hall, the Vernay-Faunthorpe Hall of South Asiatic Mammals and the Primate Hall. With the exception of the last-mentioned, all of these halls are habitat group halls, the largest part of their exhibitions consisting of lifelike scenes depicting some actual locality within the animal's range. The Primate Hall has a series of habitat groups but also contains a number of single specimens in wall cases. The Hall of Ocean Life contains the sea-mammal exhibitions, but is not entirely a mammal hall. In the corridors of the Roosevelt Memorial Building is an exhibit of Mammals of New York State. Two of the large habitat groups on the first floor of this building, the "Adirondack Forest Group" and the "Elkhorn Ranch Group," show scenes associated with the life and work of Theodore Roosevelt and feature mammals. The Elkhorn group includes some prong-horn antelopes while the Adirondack group shows white-tailed deer.

The Akeley African Hall is the largest habitat-group hall in the Museum and perhaps the finest of such halls anywhere in the world. It contains twenty-eight large groups of African mammals including in the center, a herd of nine African elephants. It is a memorial to the very man who worked almost a third of his life endeavoring to bring it into reality. It represents Akeley's dream come true for it literally is a "piece of Africa" brought to the New World and preserved for all to see, enjoy and study. One cannot tell the story of this hall without telling the story of the man who conceived it. Nor could any biography of Akeley be complete without the story of the hall that bears his name.

Akeley was not a mammalogist, but an artist and preparator. His principal medium of artistic expression was clay, but he was a versatile and talented man. He received his early training in the famous Ward's Natural History Establishment at Rochester, New York, one of the pioneer taxidermy studios in America which still prepares exhibition material for natural history museums, and has trained, in its long history, many skilled men, including Frederick A. Lucas, one-time director of the American Museum of Natural History, and other museum workers here and in many of the leading museums in the country.

Akeley's talents were many. Artistic, inventive, deeply sensitive to nature, a fine naturalist and an explorer, his chief and enduring passion was for artistic taxidermy and his consuming purpose in the latter third of his life was to create in America an African Hall which through the magic of taxidermy of the calibre to which he had raised it, would bring Africa to the American people.

"Akeley loved Africa," wrote Daniel E. Pomeroy, Museum Trustee and Chairman of the African Hall Committee. "He knew the primitive Africa, and he saw it being destroyed. It thus became the single purpose of his life to preserve and portray Africa for posterity. He dedicated his life to the task of bringing Africa to America."

Akeley's love for Africa was born when he made his initial trip to the Somaliland in 1896 under the auspices of the Milwaukee Public Museum to which he had gone following his apprenticeship at Ward's. In Africa with Dr. Daniel G. Elliott it was as though a new world had opened up to him. The beauty, the wonder, the vast unspoiled forests and plains, the teeming wild life of the world's last great stronghold of nature fired Akeley's imagination and provided him with a dream that he was to cherish and build upon for the rest of his life. "I was so bewitched by the beauty and splendor of Africa," he wrote, "that it seemed to me inevitable that I would not immediately return."

He did return to Africa, again and again, finally to be stricken ill and to die and be buried there in 1926, but it was not until nine years after his first trip that he was to see Africa a second time. The result of his trip in 1905 was his dramatic group, "The Fighting Bulls," which graces the entrance hall of the Chicago Museum of Natural History. He was tremendously interested in elephants and to him they held a fascination not equalled by any other animal.

It was his proposal to exhibit a group of elephants in ample space that brought him to the American Museum of Natural History. The Museum accepted his proposal and once again in 1909, he returned to his beloved Africa, this time as a representative of the American Museum of Natural History. His plan to collect the specimens for a statuesque group of four elephants took him to the slopes of Mt. Kenya and to Uganda. The specimens which he brought back he mounted and they have become incorporated into the larger group in the present Akeley Hall. They are among the few African mammals in the African Hall which were actually mounted by Akeley himself, most of them having been mounted by Dr. James L. Clark, Robert Rockwell and John Hope. The hall as a whole was assembled under the direction of Dr. Clark, then chief of the Museum's Department of Preparation and of Dr. Harold E. Anthony, Deputy Director and Chairman of the Department of Mammals.

Akeley did not live to see his dream come true. Dying in Africa in 1926, he was robbed even of the thrill of working on the groups in the hall. He had laid his ground work well however, and the hall in its finished state still bears his mark.

Some of his closest associates and friends feel that his death in 1926 was in part due to the lasting effects of the very severe injuries he received in 1909, on his first trip to Africa for the American Museum. It was on this trip that he was surprised by a large bull elephant. His gun missed fire when he attempted to shoot the charging tusker. The next thing he knew the elephant's tusk was right at his chest. He then did what he had mentally rehearsed as the best move in such an encounter. He grabbed the tusk in his left hand, took hold of the other tusk with his right, and swung himself in between them as the elephant began to push him down on his back. The elephant bore down and as its tusks penetrated the ground on either side of the helpless man, the curled-up trunk crushed his chest. The elephant then struck Akeley with his trunk, breaking his nose and opening his cheek, exposing his teeth. He lost consciousness at this point, and the elephant, leaving his motionless form, turned away to chase his terrified native boys. Later he learned that a minor miracle however had saved his life, when he discovered that the tusks of the elephant had met with an obstruction in the ground and had not been able to penetrate deeply enough to bring the full force of its trunk to bear against his body. Had the tusks not been blocked in this manner he would have been crushed to death. As it was, he lay unconscious for several hours. The natives, thinking him dead, would not touch him until he had regained consciousness. For three months he was in bed, for his chest had been badly damaged and several of his ribs had punctured his lungs. With his characteristic philosophical approach, he said afterwards that this enforced idleness gave him the opportunity to think out his plans for African Hall.

It was also on this expedition that he began to work out the idea for one of his inventions. He had felt the need for a motion picture camera that one could aim as one does a pistol. If he was to create life-like displays of African animals in their native haunts he needed good motion pictures of the living animals in action. Such pictures when studied later in the Museum would give the artists first hand reference material of a type which the still camera and the brush or pencil could not provide. Back in New York he began to experiment with designs for a camera that could easily be moved about and kept trained on its subject. Engineers and mechanics were skeptical and unhelpful. After much work he completed a model which looked more like a machine gun than a camera. It has been said that when a young lieutenant in World War I was trying out an Akeley camera on the front lines and was setting up this formidable-looking device, seven Germans who had just come into view, threw up their hands in surrender.

Akeley's camera has proved its value time and time again on expeditions. With modifications it made the taking of newsreels depicting scenes of action a relatively easy task and the Hollywood version of the Akeley camera meant that no longer would the actors have to parade in front of the camera but that the camera could be turned to follow them. Akeley did not develop his camera for any commercial use. He had resolved when he returned from Africa in 1911 that he would not go back again until he had produced the kind of camera he needed—one devised specifically to help him bring Africa to America.

World War I kept him from returning to Africa as soon as his camera was ready. It was while he was waiting that he worked out his cement gun, used to make the Planetarium Dome and in the building of concrete ships. He devised it first as a means of spraying liquid cement upon vertical or inclined surfaces of the manikins upon which the animal skins are mounted in the Akeley technique of taxidermy.

Akeley returned to Africa in 1921 and this time he went to the mountain paradise of Kivu to get material for a Gorilla Group. In 1926 he was back in Africa once more, working at this same spot when he was stricken with what proved to be his last illness. He was buried on the slopes of Mt. Mikeno in a tomb cut out of solid volcanic rock, to rest eternally in the land he loved on the side of a mountain in its most beautiful spot.

When in 1926 he visited some areas he had last seen in 1911 he was shocked to see a complete wasteland instead of a countryside swarming with game. It was on this occasion that he wrote to the Director of the Museum, the late Dr. George H. Sherwood, "I have not appreciated the carrying on of the African Hall, if it ever is to be done, as I now do after this painful revelation. The old conditions, the story of which we want to tell, are now gone, and in another decade the men who knew them will all be gone."

That the Akeley Memorial African Hall stands today in all its completed magnificence is a tribute not only to the man whose vision led to its development but to the group of specialists whom Akeley had trained and who had served him in the field—Rockwell, Raddatz, Leigh, Jansson and others like Albert Butler, Dudley Blakeley, John Hope, U. Narahara, George Petersen, Fred Mason and Joseph Guerry. These men working under the talented direction of Dr. James L. Clark, whose long association with Akeley made him so well fitted for the task, and the mammalogists who, working under the direction of Dr. Harold E. Anthony, assured the scientific accuracy of the finished groups and who in several instances went to Africa to collect specimens, all deserve credit for the notable achievement. Last, but not least, the hall is a tribute to the loyal friends of the Museum, Trustees, and members whose financial assistance and efforts to promote support of the venture made the dream of Carl

Akeley come true and by so doing, bequeathed to posterity a wondrous gift, the beauty of original Africa, preserved for generations to come.

The main floor of Akeley Hall was opened to the public in 1936. Its main entrance now leads into the large hall of the Theodore Roosevelt Memorial. On entering one passes under the appropriate panels of murals in the Roosevelt Hall which tell the story of Theodore Roosevelt's interest in Africa. Flanking the entrance to Akeley Hall are two bronzes of African natives done by Malvina Hoffman. Incidentally, the bronze groups of the lion hunters and the hunted lions, one of Akeley's masterpieces of sculpture, can be seen in the Roosevelt Memorial Hall. As the visitor enters Akeley Hall he is confronted by a herd of elephants which dominate the hall. These massive creatures are shown in a state of alarm. The great bull elephant is shown with his long trunk extended testing the air for scent, the young are huddled close to their mothers' flanks and in the rear a younger bull has wheeled about to cover the herd from possible attack from the rear.

The first group to the right is the "Water Hole" group. This scene is a typical animal spa along the Guaso Nyiro River in eastern Africa. Long rainless seasons dry up the streams and the animals resort to these natural shallow wells. Within this group are the reticulated giraffe, species of antelope, and the Grevy zebra. In the background are baboons, elephants, giant eland, and other animals of the dry plains. The Mountain Nyala Group is set in the hilly country of Abyssinia. This handsome creature has been called the "Queen of Sheba's Antelope." The African buffalo in the next group are shown emerging from a marsh along the Tana River in

Kenya. These seemingly docile plant eaters have the reputation of being among the most dangerous of African big game to hunt. Quite often, they will circle the hunter and come up behind him. The tall grass in the marshes they frequent hides them from view and sometimes the buffalo has turned the tables on the hunter and made him the hunted instead. Often the animals, especially the cows, will charge with little or no provocation and their massive horns make them extremely dangerous.

The African Lion Group shows a family of these enormous cats resting under the shade of a tree on the great plains of East Africa. There was a time when a good photograph of these beasts in the wild was considered a rare trophy but today, since the advent of motor cars, and the setting aside of large game preserves in Africa, such as Kruger Park, one can drive right up and "shoot" them with a camera with little difficulty. Generally speaking, lions do not molest man unless attacked. They feed on zebra, wildebeest, hartebeest and other antelope of the grassy plains. Their tawny color, as the Museum exhibit demonstrates, is excellent camouflage against the dried grass of the plains. This helps the lion to secure its prey by stalking, moving slowly and carefully and then covering the last few yards in a swift charge ending in a leap.

Among the most strikingly handsome of the African antelope are the bongo. The Bongo Group shows a pair of these vividly striped deep-forest antelope in a bamboo forest on the slopes of the Aberdare Mountains in Kenya. Their coats of rich brown are enlivened by white vertical stripes on the sides and by a crescent-shaped white collar. In the Museum group two bongos are in the act of disturbing the slumbers of another typical forest dweller, the giant forest hog. The surprise is mutual and forebodes no conflict as the bongos are vegetarians like all the antelope.

The largest of all the antelopes is the giant eland, two of which are shown in the Museum group. There are no deer in Africa but there is a wide variety of antelope. Unlike deer, the antelope do not drop their horns or antlers each year but the horns of the antelope continue to grow throughout the lifetime of the animal. The giant eland, the size of a domestic ox, is restricted to small areas in Senegal and a region between the White Nile and Bahr-el-Ghazal. They seem to prefer dry plains of thorny scrub. Being browsers rather than grazers they feed upon the leaves of trees and shrubs. Their greatest enemy is man but the region they inhabit is also sleeping-sickness country and this has helped to keep the hunters out. They are shy and wary animals and too strong to be successfully stalked by lions. Unless a lion can get very close and reach the antelope's head with the first rush, the giant eland is strong enough to shake it off and make its escape.

A large corner group shows a scene along the Upper Nile and includes some of the antelope that frequent the shores of this river, along with waterbuck, kob, Nile lechwe, tiang, sitatunga and roan, a "river horse" or hippopotamus, and crocodiles.

In the next corner the East African Plains Group shows the vast teeming herds of game that characterize this area in Africa as the last remaining place on earth where such large herds of game still exist. The bulk of these herds are composed of antelope such as the wildebeest, hartebeest, and gazelles—Robert's gazelle, for instance, and the graceful little Thomson's gazelles. Interspersed with the antelope are zebras, giraffe, and other African plains animals. The longest horns on any of the African antelope are in the Greater Koodoo Group. The great sweeping horns of the specimens in the Giant Sable Antelope Group are almost as long but the records show that the spiral horns of the greater koodoo are a few inches longer than those of the giant sable antelope. The latter is a spectacular looking animal. Its upstanding carriage, its impressive curving horns and its general appearance of alertness give it a look of vigor and watchfulness. Unfortunately this grand creature is on its way to extinction.

The gemsbok is the largest of the Oryx, a group of antelope with long, relatively straight horns. The Gemsbok Group shows these handsome mammals in their typical desert habitat on the Kalahari Desert in southwestern Africa.

The Okapi Group shows one of the larger mammals that was practically unknown to men of science before 1900. It is strange that such a good-sized animal should have escaped scientific detection until so late a date, but when one considers that it inhabits the dense jungle of the Belgium Congo, a region very seldom penetrated by white men before that time, one can understand how this strange, forest-dwelling relative of the giraffe escaped notice for so long. It seems to partake of the features of several different creatures, the zebra, the antelope and the giraffe. It has striped markings on its legs that led one early naturalist who examined a piece of the skin from an okapi's leg to surmise that it might be some kind of a zebra. Its body resembles that of the larger antelope but its long neck and its prehensile or grasping tongue betray its kinship to the long-legged, long-necked giraffe.

The Gorilla Group completes the array of splendid habitat groups on the main floor of the Akeley Hall. This group shows the mountain gorilla which lives only in the eastern part of the Belgium Congo in Central Africa. There is a lowland race, the West Coast gorillas from which the mountain gorilla differs in cranial characters, thicker and darker fur and shorter arms and longer legs. Gorillas are powerful animals and dangerous if crowded too close but they are vegetarians, and generally speaking are not aggressive. Their favorite food is the wild celery shown growing in great profusion in the group. They are not numerous and were it not for the fact that they are protected by law throughout both ranges and are able to enjoy perfect sanctuary in the *Parc National Albert,* they would soon vanish at the hands of unscrupulous hunters and the stalking leopards who kill the young gorillas which may stray from their older and stronger companions.

The mezzanine of Akeley Hall presents another series of fourteen such habitat groups which are not quite so large as the downstairs groups but present as revealing a picture of African wild life. These groups show in their natural habitat, such African mammals as the agile klipspringers, the sleek and speedy cheetah or hunting leopards, chimpanzees, the lesser koodoo, the mandrill baboons, the white and the black rhinoceroses, the hunting dogs, hyenas, leopards, colobus or horse-tailed monkeys, the springbok, blesbok, black wildebeest, and the wart hogs.

The Vernay-Faunthorpe Hall of South Asiatic Mammals is also on the second floor of the Museum. This finest and most complete exhibition of the larger South Asiatic mammals to be found anywhere, is the result of a series of six expeditions into India, Burma and Siam from 1922 to 1928. Mr. Arthur S. Vernay and Colonel J. C. Faunthorpe gathered and donated this collection to the Museum and made its transformation into realistic and beautiful habitat groups possible.

The centerpiece in this hall, as in Akeley Hall, shows elephants, but these are the slightly smaller Indian elephants, the kind most people are familiar with because they are the more easily tamed creatures which one sees in circuses and carnivals. They do not possess ears as large as those of the African species and they have a higher forehead, and an arched back.

This hall, recently renovated and repainted, is unusual in another respect—that of having its case moldings, its decorative plaster work, its window framings and its other woodwork made of teakwood all carved in typical Indian motifs to create an East Indian atmosphere. The hall shows some of the more characteristic Indian antelope and deer, the wild dogs of the hills, the gaur, the largest of the present day wild cattle, the water buffalo, the banting, Sumatran and Indian rhinoceroses, gibbons, tigers and the rarest of all the mammals in the exhibit, the almost extinct Asiatic lion.

This is the only museum in the world where one can see a pair of these lions on exhibit. They look very much like their African cousins but are slightly smaller in size and the males of the Asiatic species do not seem to have as heavy and as dark manes as those of the African. Their color is slightly paler than that of the African lions. At one time these lions had a wide range in northern India but are now found only on a protected reservation and their numbers have been reduced to about two hundred.

Still not entirely finished, the Hall of North American Mammals is the newest of the Museum's mammalian habitat exhibitions. This Hall replaced the old and antiquated Allen Hall of North American Mammals. More than twenty-five expeditions ranging from Mexico to Ellesmere Land north of Baffin Island and from the Atlantic to the Pacific coasts have been sent out to secure specimens and materials for its completion. When finished the hall will have twenty-nine alcoves, some very large. The background settings for many of the groups were selected in national parks and in similarly preserved areas where conditions are not expected to be changed. Thus, the mountain lion is shown in the Grand Canyon of the Colorado, the grizzly bear in Yellowstone Park and the coyote in Yosemite Park. The Virginia deer is shown in Harriman Park in New York State and the big-horn sheep in Jasper National Park in Alberta, Canada. Of course the locality for each setting had to be chosen within the typical range of the animal in the group, but, within this limitation, this great hall is a hall of geography as well as of mammalian life.

The Mule Deer Group is set against the background of the famous Devil's Tower in Wyoming while Shipwreck Rock in New Mexico is represented with the Spotted Skunk Group. Mt. McKinley, the highest peak in North America is the background for the white sheep, while the bison are shown in an historic setting in Wyoming, the place where the famous Overland Trail crossed the North Platte River. The Bison Group is the only one in the hall that depicts a scene one could not witness today. In the painted background are great herds of bison as they once existed in this country.

The dramatic Alaska Moose Group is set on the Alaskan

Kenai Peninsula. It shows two great bull-moose with their massive antlers, in desperate combat for the affections of a cow-moose that stands quietly awaiting the outcome. The antlers on the larger male have a spread of seventy-eight inches—the largest on record.

Dominating the center of the hall at one end is a group of the largest carnivorous land animal in the world today—the Alaska brown bear, which sometimes weighs as much as 1,600 pounds though it is much less ferocious than the grizzly.

The group shows two bears which have just interrupted the meal of an otter who is reluctantly returning to his fishing stream, while the bears approach the salmon he was forced to leave behind. One of the bears, shown standing upright, towers over eight feet tall. The background depicts the typical volcanic peaks of the Alaska Peninsula with the glaciers and snow fields that supply the streams clinging to the slopes and upper valleys of the mountains.

The Musk-ox Group shows a small herd of these hardy northern mammals making their way out of a blizzard and the caribou groups show the Barren Land and the Osborn caribou. The Rocky Mountain Goat Group is set in one of the American fiords along the Alaskan coast.

Not all of the North American Mammals can be shown in a hall of this type but the outstanding mammals have been selected. In addition to groups already mentioned are group exhibits of the jaguar, the Florida black bear, the timber wolf, the cottontail rabbit and jack rabbit, the eastern and western skunk, the beaver, the otter, the Canada lynx, the fisher, and the porcupine.

The habitat groups of sea-mammals such as the walrus,

the elephant seal, the manatee, the Stellers' sea lion, the fur seals, both the Townsend's and the Alaska, are in the Ocean Life Hall along with the models and skeletons and paintings of whales and other cetaceans. The Pacific Walrus Group, one of the largest in the Museum shows these Arctic sea-mammals at home on an ice floe.

The cetacean collection contains among other things a full-sized model of the killer whale alongside a skeleton of this fierce predatory giant porpoise, a creature capable of swallowing a fur seal or a small porpoise at a gulp. These killers of the sea have been known to attack and kill even the large baleen whales.

Large skeletons of the sperm whale, the right whale and the baleen are suspended from the ceiling along with models and skeletons of smaller cetaceans such as the narwhal, with its long tusk, the pygmy sperm whale, the white whale or beluga and many others.

The Ocean Life Hall also has series of large mural paintings by John P. Benson, the noted marine painter. Four of these murals, "The Chase," "The Attack," "Towing the Carcass" and "Trying Out,"—show scenes typical of American sperm whaling. Other murals show whales including a sperm whale attacking a giant squid and one of a pack of killer whales attacking a gray whale.

The Giant Panda Group and the Siberian Tiger Group are on the second floor and are the only groups of North Asiatic Mammals now on exhibit. They are forerunners of a proposed hall of North Asiatic Mammals which the Mammal Department hopes to be able to develop in the future.

The many, varied and beautifully constructed habitat groups in the Museum's halls that deal with vertebrate life have made the Museum exhibitions a delight to the millions of visitors who have seen them. They represent only one type of Museum exhibit however, and the Museum's exhibition program has devised other effective presentations of material. The large habitat groups represent an expense sometimes far beyond that imagined by the casual visitor. Some of these groups have cost their donors over $40,000 and a large hall of groups such as the Akeley costs over $1,000,000. The care and patience and long hard detailed work that goes into their creation is sometimes overlooked by the visitor who takes in their beauty in a glance. This story will be amplified in the chapter "Behind the Scenes."

CHAPTER SIX

Man before History

THE STORY of Man as displayed in the exhibition halls of the American Museum of Natural History is a long, elaborate and fascinating tale. It is a story that spans more than a million years and draws its main characters from all over the world. Actually, all of the exhibits at the Museum that deal with life, even of the most primitive and tiny kinds, are part of the great and still unfinished saga of evolution of which man's relatively short and recent career is but the latest chapter.

The Department of Anthropology, now headed by Dr. Harry Shapiro, and staffed with a coterie of capable ethnologists, archaeologists, and physical anthropologists, has as its field of investigation the study of mankind, past and present. Organized in 1873, the Anthropology Department did not receive its real start in life until 1895 when it came under the able leadership of Dr. F. W. Putnam. Up to that time the department had acquired by gift and by purchase a number of ethnological and archaeological collections but it had not undertaken any expeditions worthy of the name, nor had its caretakers installed anything resembling a program of study activities. In the early years, until 1891, it was under the care of Professor Bickmore. From 1892 to 1894, Mr. James Terry was in charge. Finally the Department was reorganized in 1895 and Putnam and his assistants, Franz Boas and Marshall Saville started it on its new career. With a trio of trained, capable and enthusiastic anthropologists the department began the remarkable work which was to make it one of the outstanding departments of its type anywhere in the world. Such scholars of this important branch of natural science as Waldemar Jochelson, Waldemar Bogoras, H. I. Smith, Alex Axelrod, Livingston Farrand, James Teit, George Hunt, Philip Jacobsen, Roland Dixon, Berthold Laufer, Pliny Goddard, Robert Lowie, Herbert Spinden, Louis Sullivan, John Swanton, George Vaillant, Wendell Bennett, Nels Nelson and Clark Wissler were to participate in its development and growth in the Museum. Dr. Clark Wissler directed the department from 1902 until his retirement in 1942. Today such scientists as Margaret Mead, Junius Bird, Gordon Ekholm, Bella Weitzner, James Ford and Harry Tschopik, Jr. continue the fine work. Clarence Hay is both a Trustee of the institution and also an important participant in the Anthropology Department's research program.

Putnam inaugurated a brilliant series of field researches and collecting trips. President Morris K. Jesup was a tower of strength for the department and provided the means and the encouragement for many expeditions. Others who helped the department in these busy years were the Duke of Loubat, Henry Villard, C. P. Huntington, Theodore Cooper, William R. Warren and B. Talbot Hyde. Anthropologists Carl Lumholtz and Ales Hrdlicka also went on expeditions to Mexico for the reorganized department.

One begins the story of "Man Before History" in the Museum by first visiting the Hall of the Natural History of Man on the third floor. It is the first part of this hall, where exhibits furnish an "Introduction to Human and Comparative Anatomy," that the anatomical story of man's place among the vertebrates begins. The plan of this hall was worked out by Dr. William K. Gregory and it tends to follow the lines of his book, "Our Face from Fish to Man." The exhibit begins with a treatment of *Man in His Cosmic Aspect*. Here Man is conceived as a living engine deriving energy from the sun, both directly and indirectly, for that which he receives from plant and animal food is also stored solar energy. Two life-sized dissection models of the human form are shown, one with just the superficial layers of skin and muscle over the body cavity removed and the other showing the internal organs. Anatomy students, students in biology, and student nurses frequently use this exhibit. Not all who have been observed studying these models have been students. On one occasion a voluble matron was observed delivering a veritable discourse to a friend. She was using the deep dissection model as illustration of her lecture. A Museum eavesdropper soon learned that she was telling her friend about her operation and using the excellent three-dimensional material in the case in front of her to enhance her vivid description. From her forceful assurance and her apparent familiarity with the model, it appeared that this was not the first time she had used this technique. She did so love to talk about it!

It appears to be the consensus of scientific opinion that our pre-human ancestors managed to become human only after millions of years of long, slow progress from lower to higher stages of life. Despite the fact that enormous numbers and varieties of living forms existed in past ages, the fossil-diggers have unearthed but a relatively small number of specimens that represent forms which appear to lie in or near the direct line of ascent from fish to man. Dr. Gregory has set up in this hall a series of ten fossil and primitive animals and not so primitive primates which help to make clear in a general way the major steps or advances which ultimately led to man.

Cheirolepis, a fossil fish from the Devonian of Scotland, represents Stage One in this series. Looking somewhat like a trout, this fish breathed by gills in the normal way but its tail resembled that of a shark. Its progress through the water was by a wriggling movement produced by the regularly arranged muscle flakes along each side of the body. Instead of a bony spine it had a notochord, an elastic rod. This notochord appears in all the higher vertebrates, including man, in embryonic stages. The fins of *Cheirolepis* were folds of skin which served as keels and rudders.

The second step is represented in this exhibit by another fossil fish called *Eusthenopteron,* which lived about 300 million years ago in the Upper Devonian Age. It still had gills but it may also have possessed an air-sac or lung. Instead of fins however, it had two pairs of paddle-like appendages. These paddles correspond to the fore and hind limbs of the four-footed animals.

Stage Three is represented by the remains of the oldest

known type of quadruped—a fossil amphibian called *Diplovertebron* which lived in the Carboniferous Age. The limbs are far advanced over the paddles of the *Cheirolepis* and there are five digits on each of the hands and feet. If the mathematical historians are correct in their theory that man evolved the use of tens as a basis for counting because this number represented the total number of digits of both hands or of both feet, then we might go a step further and say that the decimal system had its origin back when *Diplovertebron* was born, for it was the first creature to have five toes on each foot.

The fourth major step in the evolution of the vertebrate skeleton is represented by the primitive reptile *Seymouria* which lived in lower Permian times in Texas, about 235 million years ago. On the whole, *Seymouria's* skeleton is not very different from the older *Diplovertebron* but its limbs were larger and better developed and held the animal's body higher off the ground.

Step Five is a mammal-like reptile named *Cynognathus* from the Upper Triassic of South Africa. Its limbs were much better fitted for running and certain features of the skull, backbone and limbs resemble those of mammals.

Stage Six is represented by a modern or recent mammal, the opossum. It has been labeled as an archaic mammal in this exhibit, for although it is common today its skeleton retains many of the prominent characteristics of the skeletons of older fossil mammals. The opossum has five-toed grasping hands and feet and uses these to climb about in trees. Its skull, teeth and brains are of a low type. Thus it may be called a "living fossil."

For the seventh stage in this progression of vertebrate life the exhibit presents another prehistoric creature, the *Notharctus*, which represents the lower limits of the order of Primates. *Notharctus* was adapted to a tree-life. Its eyes and brains were much larger than any of the types of vertebrates that had existed before it.

The eighth step is portrayed by the modern gibbon, an East Asiatic ape believed to be a descendent of the first of the tail-less or man-like apes although it is a tree-dweller itself. When on the ground, however, it walks normally on its hind legs and is the only existing man-like ape to do so. Its skeleton is man-like in many ways but its arms are excessively long so that the knuckles touch the ground even when it walks on its hind feet.

Stage Nine is represented by two modern primates, the chimpanzee and the gorilla. These apes retain the more important characteristics of the fossil apes of India and South Africa, which in turn have many of the characteristics of the oldest known early men. Their brains are much more developed than those of any of the lower animals and in intelligence they more nearly resemble man than any other animal types.

The final stage presented by Dr. Gregory's series is, of course, man. The exhibition shows clearly that the human skeleton is built upon the same general plan as that of the gorilla, chimpanzee and gibbon. In man the backbone, pelvis and limbs are modified so that he can walk erect on his hind legs and thus have his arms or forelegs free to do with as he wishes. His brain is much larger and is more highly developed than is that of the apes.

The exhibits in this hall present a family tree of the primates and such comparative sequences as the series of reconstructions depicting the evolution of the face from fish to man. A chart and an exhibit illustrate the evidences of evolution from an embryological point of view. In the exhibit devoted to the brain and its evolution again a large wall chart compares the brains of man with those of the successive lower forms of animal life.

The Hall of the Natural History of Man consists mostly of planned displays featuring comparative anatomy and although it is a good starting place for the story of prehistoric man, the visitor will find that the Hall of the Age of Man on the fourth floor of the Museum will show him much more of the fossil evidence of early man and of the animals that were contemporaries of his. The Age of Man goes back about one million years and the fossils in this hall represent this span of time.

Most of them are fossils of the animal contemporaries of the oldest humans. Fossil remains of prehistoric man, especially of the earlier types, are extremely rare and the Museum's exhibit represents by actual specimen or by cast the bulk of the fossil material of early man which is known to exist. Most of the original bones and skulls are in European museums and the Museum's paleontologists and anthropologists have had to obtain casts of these rare specimens.

The Age of Man dates back to the beginning of the Great Ice Age or Glacial Period. This period, known as the Pleistocene, saw vast areas of northern Europe and northern North America buried deep under huge continental glaciers which advanced and retreated and advanced again. We are now living between two ice ages. The slow retreat of the ice is still going on in the north and it is thought that after this retreat slows down to a standstill an advance of the ice may take place again. This is a slow process and it has been estimated that many thousands of years will have to elapse before the cycle can complete itself.

Man may have had his beginnings in the Age of Mammals but some authorities think that the Ice Age helped to speed up his development. The colder climate of the northern regions in the Pleistocene put pressure on the mammals. The growing season for plants in the north became shorter and the plant-eating mammals of this region had to develop feeding habits that would enable them to survive during the winter months. The competition that took place must have weeded out the slower-witted stocks and placed intelligence at a premium. Thus, although the human prototypes must have separated from the rest of the primate stock near the middle of the Age of Mammals, it took millions of years before the special endowments that characterize mankind were developed to the stage where reason, memory, and speech combined to give mankind a great advantage over the rest of the mammals. Far outnumbered by his mammalian contemporaries, fighting by wits instead of sheer brawn, early man slowly but surely began to dominate the animal world and wrested his superiority in the animal kingdom from larger and more powerful but less intelligent mammals that had held that position.

The earth has known ice ages even before the Pleistocene. There is evidence that at least four times before the Age of Man, large areas of the earth were subjected to a polar type of climate. Even today, we can find rivers of compressed ice and snow in the Alps, Rocky Mountains and high mountain areas elsewhere. Greenland and Antarctica are great land masses which are still buried under thousands of feet of the continental type of glacier. Geologists, including those on the Museum's staff such as Dr. Chester A. Reeds, have made studies of the layers of clay and mud which have been deposited in glacial lakes in the past. These layers or *varves* furnish a means of timing the duration of the glaciers and many studies have indicated that at least 35,000 years have passed since the continental glaciers of the Age of Man began their last retreat. According to Antevs, the present climate of the earth was established about 10,000 years ago and this may be taken as the beginning of recent time.

As the ice sheets advanced and retreated so did the meadows and forests, and so did the mammalian life in the northern half of the world. With the closing phases of the Age of Mammals came the extinction of many superb forms of mammals, a direct or indirect result of the severe climatic changes which came with the early part of the Ice Age. The Hall of the Age of Man was planned to show what actually is known of man, his mammalian contemporaries and his environment during the million or more years of geologic time when he rose to a position of dominance.

The center of the hall is devoted to a series of floor cases which present chronologically the older races of mankind. The evidence presented consists of fossil remains or casts of fossil remains, fragments of the handiwork of these early humans and restorations of their features. It is difficult to determine the age of these remains. The archaeologists have created a sort of time sequence which has become popularly known as the Stone Ages. This series of ages begins with the Eolithic or Dawn Stone Age, passes into the Old Stone Age, then to the New Stone Age. Between the Old Stone Age and the New Stone Age is a period known as the Transitional. After the New Stone Age, the Ages of Metal begin with the Copper Age, followed by the Bronze and Iron Ages. The Historic Period as differentiated from the Prehistoric, has its beginnings about 5,000 B.C. which also marks the beginning of the first of the metal ages. This time system obviously is based not upon man himself but upon the samples of his handiwork. The Hall of Stone Age Culture on the second floor of the Museum presents this archaeological story in considerable detail but the Hall of the Age of Man exhibits are more concerned with the actual organic clues to man's past.

The study of fossil man is a very recent one. It was natural that the first fossil finds were likely to be those resting near inhabited places and those of more recent early man which would exist in greater profusion than those of the very earliest forms. Remains of Neanderthal man were found as early as 1700 but the first important specimen was discovered near Gibraltar in 1848. Even at this time its significance was not understood. In 1856 some fragments were found in a cave deposit in the Neander Valley, near Dusseldorf, Germany. On the basis of this find William King created the name *Homo neanderthalis* in 1864, the German *thal* or *tal* meaning valley. Later, in 1886 some skeletons of the "Neander Valley" men were dug up near Spy, Belgium. Numerous flint implements, fossil remains of mammoth, wooly rhinoceros and other extinct mammals at last showed the archaeologists that the Neanderthal race was a distinctive and archaic species of mankind, and that *Homo sapiens* was not the sole representative of the Hominidae to occupy the earth.

Since that time almost a hundred more specimens of Neanderthal man have been found in widely scattered localities from the Crimea to Germany and Spain. Neanderthal is not the most primitive or oldest of the early races of man to be seen in the Museum's exhibit, but it is probably one of the most widely known and for years it was thought of by the layman as the typical "Cave Man."

One of the lowliest if not the lowliest of the early forerunners of present man is the "Java Ape-Man" or *Pithecanthropus erectus*. Found in an early Pleistocene deposit near Trinil, Java by Dr. Eugene Dubois in 1891 and 1892, the "Java Ape-Man" is also known as "Trinil Man."

Dubois had gone to the Netherlands East Indies with the stated purpose of discovering an early form of man. The parts that were found consisted of the upper part of the cranium or brain case, a femur or thigh bone and three teeth plus a fragment of the lower jaw bone. The teeth and the jaw bone fragment were not found together nor were they found with the piece of skull and the femur. Not all scientists of this field of study agree that these fragmentary pieces all belong to the same individual nor that they are all parts of a human form. Some hold that the skull-cap and the femur are from different individuals and that the teeth are those of a fossil orang-utan.

Dubois called his find *Pithecanthropus erectus* ("ape-man of erect posture") because the skull is very ape-like and because the femur shows the creature had an upright posture. It has been long debated whether this was a highly developed ape or a primitive man.

The Museum exhibit on *Pithecanthropus erectus* contains casts of the fossil remains, a brain cast made by using the skull-cap as a mold, and a restoration of the head made by

Professor J. H. McGregor. The age of this creature has been estimated at not less than 500,000 years. The brain capacity was only 900 cubic centimeters, which, although it far exceeds that of the modern great apes, is far below that of living races of man.

In 1937 more remains of early man were found near Trinil by Dr. G. H. R. von Koenigswald. These bits consisted of another cranium and part of a lower jaw with four teeth. The brain capacity of the individual appears to be only about 750 cubic centimeters, half that of modern man. It is thought that these fossils are those of a creature like *Pithecanthropus*. Another skull found near Modjokerto, Java seems to be that of an infant. Dr. von Koenigswald has suggested that this may well be an infant *Pithecanthropus*.

In the Hall of the Age of Man, is an exhibit devoted to the "Peking Man." At about the turn of the century, Dr. Max Schlosser bought some "dragon's teeth" in an apothecary's shop in Peking. The Chinese have long used fossil mammal teeth as a pharmaceutical. Schlosser found a tooth which seemed to be human in his assortment of "dragon's bones." This led eventually to the consideration of China as a possible source of the remains of early man.

In 1923 some fossil human teeth were found near Peking in a cave deposit and eventually more teeth, broken jaw bones and some crania were found. The National Geological Survey of China has uncovered fossil specimens representing at least 27 individuals of this prehistoric race of man which has been named *Sinanthropus*. The skull seems more highly developed than that of *Pithecanthropus erectus* and the brain capacity averages about 1,035 cubic centimeters, or nearly fifteen percent larger. In addition crude tools of stone, bone and antler indicate that "Peking Man" understood the making and using of tools and of fire.

The exhibit contains reconstructions of the skull of "Peking Man" made by Lucile Swan who worked under the direction of Professor Franz Weidenreich, one of the outstanding students of prehistoric man.

In the early 1940's Dr. Weidenreich worked at the Museum on the restoration of the skull known as Pithecanthropus IV. This skull from Java, procured by his younger colleague, Von Koenigswald had been badly crushed and in making his restoration from casts, Weidenreich was faced with a task of considerable difficulty.

Skull I was the original *Pithecanthropus* which Dubois had found in Java; Skulls II, III and IV are more recent finds. There seemed to be quite a physical resemblance between the *Pithecanthropus* and the *Sinanthropus* races. Yet, Weidenreich says of the skull of Pithecanthropus IV, "There is no doubt that Pithecanthropus Skull IV represents the most primitive type of fossil man ever found." The jaw of this specimen was massive, like that of the "Heidelberg Man" whose lower jaw bone and teeth were found in 1907 at Mauer, near Heidelberg, Germany, in a commercial sand pit.

There seems no doubt now that the two oldest known races of man were the *Pithecanthropus* and the *Sinanthropus,* both these types being much earlier than Heidelberg, Piltdown or any of the types of the later Neanderthals and Cro-Magnons. According to Weidenreich, *Pithecanthropus* "shows more human traits than simian, and further exploration may reveal that he was a skillful hunter acquainted with fire like his cousin, Peking Man. Whether he had a language, a religion, or family organization, science cannot yet say."

The fragmentary remains of "Piltdown Man," discovered by Charles Dawson and Smith Woodward in 1908, 1911, and 1912, included pieces of the cranium, and a piece of lower jaw bone with some teeth. Along with a worked flint they were found in a gravel bed near Piltdown Common, Sussex, in southern England and given the scientific name of *Eoanthropus dawsoni,* or "Dawson's Dawn-man." The skull appears to be unusually thick but this primitive characteristic is somewhat offset by the fact that the brain capacity of Piltdown, between 1,200 and 1,400 cubic centimeters, was greater than that of either *Pithecanthropus* or of "Peking Man."

The first "cave man" that we really know something about was "Neanderthal Man." The large number of remains found and the associated tools and weapons discovered with the bones and the fossil animal remains found next to some of the fossils of Neanderthal make it possible to gain a few clues to the mode of life of this primitive European. Knight's mural on the Neanderthal shows a small group of these Old Stone Age people squatting around a fire at the entrance of a hillside cave in France.

Their crudely made tools and implements included handaxes, scrapers, points and some bone tools for skinning and flint-flaking. Nothing resembling ornamentation has been found, so we have no indication of any Neanderthal artistic or aesthetic sensibility. The charred remnants of hearth fires have been uncovered. No one knows when man first learned how to harness the power of fire but we know that "Neanderthal Man," and perhaps, "Peking Man" had done so. The brain of the average *Homo neanderthalis* was nearly as large as that of the average modern European, but its convolutions seem to have been more gross and simple, and the grey matter thus less developed than in the brain of man today.

The most recent of the early races of man in Europe was the creature called "Cro-Magnon Man." These Upper Old Stone Age people are thought to have been a group that entered Europe from the east and probably drove out the Neanderthals. The Cro-Magnons were essentially like modern man in anatomical structure and their skulls reveal advanced stages of developments. Hunters, painters and sculptors they were a race far superior to any of their precursors. Tall, with brains larger than those of the average modern Europeans, these intelligent and talented people were responsible for many of the fine cave paintings and pieces of cave sculpture found in France and Spain. Mr. Knight's mural of the Cro-Magnon shows a group of artists at work in the cavern of Font-de-Gaume at Dordogne, France where the easily recognized animal figures being drawn and crayoned on the cave wall have given us a clear idea of the prehistoric animals these people knew. We see mammoths, pre-historic bison, horses, deer and other creatures of the late glacial period.

Following Cro-Magnon there were several groups of Old Stone Age people who lived in or near the transitional period between the Old Stone Age and the New Stone Age. They are on display along with some of their tools. Knight's third mural shows a group of stag hunters of the Neolithic or New Stone Age. Some of the cultural advances they had made over their old stone age cousins are evidenced by houses made of mud and sticks, wolf-like dogs which were tamed and used in the hunt and a new weapon, the bow and arrow, destined to be the chief long range offensive weapon of man for thousands of years to come.

Another set of murals in the Hall of the Age of Man show animals that flourished during that time: "Midsummer on the Ohio River in Late Glacial Times," "Autumn in Northern New Jersey During Late Glacial Times," "The Prehistoric Death Trap of Rancho-La-Brea, California," "A Loess Storm on the Pampas of Argentina," "Reindeer and Mammoth on the River Somme, France," and "The Woolly Rhinoceros in a Glacial Winter, Northern France."

There is an outstanding series of fossils in the Hall of the Age of Man showing the evolution of the *Proboscidea*. Two groups are shown, the mastodents, and the mammoths and elephants. The earlier forms of mastodons had four tusks, two in the lower and two in the upper jaws. The series indicates the gradual reduction in the number of teeth and the gradual increase in size. One of the specimens shown is the famous Warren Mastodon, a fossil dug up near Newburgh, New York in 1845. Other mastodon remains from the Hudson Valley and from the City of New York show that these hairy elephant-like creatures were native to the present metropolitan area. There is also a group showing a tar-pool at Rancho-La-Brea with a saber-tooth tiger, a dire wolf and a giant sloth. The sloth was first to be trapped, the tiger going to feed on him is also shown becoming mired and the dire wolf standing on some hardened asphalt overlooks the scene and one feels that his appetite will get the best of his wisdom and that he, too, will soon be caught in the sticky tar.

Nearby, a very large floor case shows some of the giant

ground sloths from South America. In this case are several large glyptodonts, gigantic prehistoric relatives of the present-day armadillo. In another part of the hall is the fossil tusk, sixteen feet long, of *Archidiskodon imperator,* dug up at Post, Texas. It is the largest fossil elephant tusk ever found, in fact the largest elephant tusk, prehistoric or modern on record.

In 1925, a cowboy was riding by a deep-cut ravine in northeastern New Mexico. Seeing some bones sticking out of the bank of the ravine he stopped, dismounted and pulled out one of the bones which he took back to the owner of the ranch who, he knew, was interested in such things. A local collector then went out to the spot and it was his report that led to the excavation of the site by the American Museum of Natural History and the Colorado Museum of Natural History. Soon the complete skeletons of a small herd of prehistoric bison came to light. With these animal bones were a number of flint dart-points of a type never seen before. It soon became evident that a group of prehistoric American hunters had killed the bison and this was the spot where they had skinned their kill. The time of this event was placed somewhere between 10,000 and 20,000 years ago and it thus became evidence of the earliest known human habitation of the western hemisphere. Thus the former belief of anthropologists that man was a very recent arrival on this continent was superseded. The skeletons of the bison and some of the flint dart-points are now in the Age of Man Hall.

Bit by bit, archaeologists are putting together the picture of early man. The future may see the uncovering of more evidence that may force the museums to revise their present displays on prehistoric man, but for the present the collection at the American Museum of Natural History represents the bulk of the evidence which the world now possesses concerning its earliest human inhabitants.

We have a great deal of material evidence on the culture of early man, particularly from the Upper Old Stone Age on to the Iron Age. The exhibit on the second floor contains thousands of bone, stone, flint and metal tools and implements as well as a number of examples of man's early ventures in artistic expression. The earliest known records of man's observations of nature are engraved drawings of animals which have been etched on reindeer antlers, ivory and limestone by prehistoric artists using sharp splinters of flint as their incising tools. These artists also sculptured animal and human figures in the round in bone and limestone. Sometimes they worked in clay, as evidenced by clay modelings found in caves in Europe. Since the collection is arranged chronologically we can see, almost step by step, man's success in mastering the art of making tools. We see man's inventive mind at work when we discover his early development of the pin, of buttons, and finally of the fibula which is nothing more nor less than a safety pin. The hunters and warriors also developed their weapons of the chase and of the battlefield. From crude hand-axes which might have been used as daggers, we see a progression of finer made blades of stone and later metal, and of spear and arrow points of stone and then of metal.

With the coming of the New Stone Age man began the domestication of animals; he discovered how to make pottery and he began to improve his agricultural skill. This latter development was to bring to a close his career as a hunter and see his transformation into a husbandman. This made the beginning of permanent settlements possible and an example of such an early settlement is to be seen in the Swiss Lake Dweller's model in the Stone Age Culture Hall. This Hall perhaps more than any other in the Museum shows how man learned to conquer his environment in the days before history.

CHAPTER SEVEN

Native Americans

AMERICA HAS always been the land of opportunity for people of the Old World. Long before Columbus started one of the world's greatest exoduses, nomadic hunters from Asia made their way into America through Alaska. The first wave of the Mongoloid migration to America probably took place during the late Pleistocene Period and may have occurred just before the extinction of the mammoth, horse, camel, sloth, and the *Bison antiquus taylori.*

The bison was the creature hunted by Folsom Man, the earliest known inhabitants of North America. The Folsom people were of a very crude hunting culture, living on bison flesh and were far less advanced than the cave painters and sculptors of the Old Stone Age in Europe. There is no reason to believe that Folsom Man was the very first of the Mongoloid migrants. However, since archaeologists are of the opinion that the first, adventurous Siberian hunters made their perilous way across Bering Strait some 25 thousand years ago and since Folsom Man seems to be dated from 10,000 to 15,000 years ago, we can be fairly safe in assuming that Folsom Man was not unlike the first of the Americans.

The Asiatic origin of the first people to populate America is borne out by the studies of the physical anthropologists, most of whom now subscribe to the theory that there are three great races of mankind; white or Caucasoid, black or Negroid, and yellow-red or Mongoloid.

There is a great similarity between the skin color of the Chinese and that of the American Indian. Both the American Indians and the Mongolians have straight, coarse black hair, compared with that of the Caucasoids which is finer, tends to be wavy and ranges from blonde to black in color, and the kinky or frizzy hair of the Negroid peoples.

Another physical characteristic common to the American Indians and the Asiatics is the "shovel-shaped" upper incisor teeth. This characteristic is rarely observed among the Negroids and almost never seen in the Europeans. Sometimes the American Indians are to be recognizable by their "high cheek bones." Indians do have unusually wide faces and this is because of the prominent zygomatic arches which they, like the Mongolians, possess. Although the Indians do not have the fold in the upper eyelid which produces the effect of a "slanting eye" they do have eyelids which tend to turn down at the inner corners as do the eyelids of many of the marginal Mongoloids. All of these physical similarities indicate that the Indians and the Mongolian peoples must have had a common ancestry.

The Mongoloids held the only overland gateway to the

New World. It is difficult to conceive of any other primitive people reaching America, except by this land and ice route. Man in the Old Stone Age did not possess any seaworthy craft that would have enabled him to make a long voyage over the open seas. As the ice retreated from Eastern Siberia it is believed that the Mongoloids moved northward and eastward and were eventually able to push their way into Alaska and from there moved southward into the main part of the North American Continent.

The Anthropology Department of the American Museum of Natural History has long been interested in the study of man in the New World. The pioneer researches in this field were made possible by a series of expeditions financed by Morris K. Jesup. Starting in 1897 these expeditions were led by such men as Franz Boas, Waldemar Jochelson, Waldemar Borgoras and Livingston Farrand. The primary problem was to determine whether or not there was any connection between the peoples of Asia and of northwest America. At the same time that Boas and his colleagues were in Alaska and Siberia, the B. T. B. Hyde Expedition was exploring the cliff houses of Utah and the ruins of Pueblo Bonito in Chaco Canyon, New Mexico. Shortly before the Jesup Expeditions to the north, Museum archaeologists were working in Mexico and Central America. Since 1894 when Dr. Adolph Bandelier worked in Peru, Ecuador, Bolivia and Chile, the American Museum's archaeologists have explored the early civilizations of South America.

Today the areas of Alaska, Middle America and the Andes are still of special interest to the Museum's archaeologists. The recent discovery of the Ipiutak culture in northern Alaska represents the earliest horizon yet to be uncovered in this area and way down at the far end of South America, Junius Bird's studies in Chile have secured evidence of early man.

It so often happens that we neglect the treasures at our doorstep and go off searching in far away places where the "grass looks greener." The American Museum did not overlook the native Americans of its own country, and in 1900 began its extensive collections of the cultural materials of the modern Indian descendents of the Mongolian migrants. The first work in the western plains area was started by Dr. A. L. Kroeber among the Arapaho. Systematic collections were made, eventually to represent practically every Plains tribe. Extensive notes, containing a wealth of information regarding a culture which has almost passed away, were also taken so the full histories of the items collected would be carefully recorded for future reference. The Indians were most cooperative and took a deep interest in the work. One old Indian said to Dr. Clark Wissler, "Now I pass in peace. You have written down our history; you have put away in a safe place the things of the old people. Our grandchildren can read and see what their ancestors did. Otherwise all would be lost. It is good that you came before it was too late."

Two areas in the Americas witnessed the development of extraordinary cultural achievements. The heights reached by the Mayan and Aztecan cultures in Central America and Mexico and the Incan culture in Peru were to rival those reached by the ancient cultures of the Valley of the Tigris and Euphrates and the Upper Nile in the Old World. Completely insulated from the Old World these great American empires were to demonstrate how far the Amerinds had advanced from their earlier Mongoloid nomadic ancestors. Two exhibition halls at the American Museum of Natural History are almost entirely devoted to presenting the stories of these two archaeologically rich New World areas. The most recently refurbished of these two halls is the Hall of Mexican and Central American Archaeology.

Completely revised by Dr. Gordon F. Ekholm and by Mr. Clarence L. Hay, the Mexican and Central American Archaeology Hall has been designed to meet two needs. On the one hand it endeavors to give a general view of the outstanding achievements of the Mayans and Aztecs in architecture, in art, in religion and in government. In addition to a general treatment of the two cultures the hall's exhibits deal specifically with cultural sequences such as that of the Valley of Mexico. The more important and spectacular treasures of the collection are in the center and in a small entrance foyer. Actual specimens of sculpture, pottery, jade and metal work as well as reconstructions of some of the more important buildings and stelae of these ancient Americans are to be found in this hall. Many of the larger pieces have been faithfully cast in molds made from the original pieces which still remain in Mexico. Because of the fine research of such outstanding archaeologists as the late Dr. George Vaillant, Clarence L. Hay and Dr. Gordon Ekholm, there is here presented the only history of the culture sequence of the Valley of Mexico to be found in any museum.

For almost 2,000 years, the Valley of Mexico was one of the most important centers of a developing Indian civilization. This great period of development and achievement was brought to an end in 1521 when Hernando Cortez conquered the Aztec capital of Tenochtitlan which stood on the site of the present Mexico City. The past four hundred years have served to bring about many changes but there is a constant reminder of the former Aztec glories to be found in the many artifacts that have been uncovered whenever building construction in Mexico City necessitates excavation and also in the fact that Mexico City has a tendency to sink into the fill that was dumped into the ancient lake over which it is built and on whose islands the Aztecs made their capital. Cortez started to fill in the swampy and marshy lake as a step in his battle plan to capture the Aztec stronghold and time and man have almost completely succeeded in obliterating all trace of the Aztec metropolis.

Among the Museum's rarest treasures are some of the priceless gold ornaments from Costa Rica. It was gold such as this which attracted the early soldiers of fortune to America. Later, this same gold was to attract pirates, and the West Indies and the Gulf of Mexico were to be the favorite cruising areas of such men as Sir Henry Morgan and Captain Kidd. Much of the gold which was taken from the Aztecs in Mexico and for that matter from the Incas in Peru found its way into European treasuries and was melted down into bullion. Perhaps some of it still rests in this form at Fort Knox. Most of the masterpieces of the Indian goldsmiths were thus lost for all time. Some pieces were so beautiful that the Spaniards sent them back to Spain to be preserved for exhibition purposes.

The discovery of Mexico has been credited to Francisco de Cordoba. Cordoba set sail from Cuba in 1517 and skirted the northern and eastern shores of Yucatan. When he landed and attempted to renew his water supply he was set upon and badly defeated in a battle with Maya Indians. Two years after he returned in defeat to Cuba, the Governor of that Island sent out the man who was to become the conqueror of Mexico, Hernando Cortez. The Aztecs were the dominant nation at the time Cortez set out, the Mayas being on the descendent. The art and religion of the Aztecs was largely built upon those of the nations which had preceded them.

The relative merits of the Mayas and the Aztecs may be said to resemble those of the two Mediterranean cultures of the Greeks and the Romans. The Mayan people, like the Greeks of old were an artistic and intellectual race who developed sculpture, painting, architecture, astronomy and other arts and sciences to a high degree. Although the Maya and the Aztec both are called nations and although they had a common religion, each was composed of a number of communities or states which were frequently at war with one another. Of course the Mayan religion was more barbaric than that of the Greeks but in both instances these two cultures idealized and beautified their deities in art and architecture. The Aztecs, like the ancient Romans, were a brusque and warlike people who built on the foundations of earlier civilizations which fell before the forces of their arms. Like the Romans, the Aztecs made their most notable contributions in the fields of organization and government.

Cortez worked his way inland toward the Aztec capital by persuading the allies of the Aztecs to join forces with him, or, when that was impossible, by defeating them in battle. Large quantities of gold were stored in Tenochtitlan for this was the time of the height of Aztec power. Some of this gold was the result of Aztec manufacture but much of it was tribute. A particularly large treasury was kept in the Montezuma's palace. Bernal Diaz, one of Cortez' men, described this treasure room in these words, "Cortez and some of his captains went in first and they saw such a number of jewels and slabs and plates of gold and other great riches that they were quite carried away and did not know what to say about such wealth. When I saw it I marvelled and, as at that time I was a youth and had never seen such riches in my life before, I took it for certain that there could not be such a store of wealth in the whole world."

Much of the gold which Cortez and his men took from the Aztecs after the imprisonment and death of Montezuma was lost on the "Sad Night" when the Spaniards made their withdrawal from the Aztec capital. The Aztecs killed many of the Spaniards, and still more of the Spaniards who fell or tumbled into the canals which threaded through the ancient city were drowned by the very weight of the gold with which they had loaded themselves. Cortez returned again to Tenochtitlan, this time to raze the city and to defeat the Aztecs.

The Museum's collection makes it possible to see a revealing segment of the Aztec world. One of the interesting relics of Aztec days is the copy of the great calendar stone, which, in its original condition was about twelve feet across and may have weighed forty tons. It must have been dragged by sheer man power over miles of swampy lake bottom before it could be placed in its original position in front of the Aztec Temple of the Sun. Some authorities felt that it was placed horizontally so that it could be used as a sacrificial altar, but a later discovery in Mexico City of what appears to be a miniature of the stone depicts it in an upright position. In any event, it was thrown down from its place by the soldiers of Cortez and for a time disappeared from sight. In 1560 it reappeared again but was once more buried by the order of the Archbishop of Mexico City who feared that its presence might incite the Indians to revert to their pagan beliefs. In 1790, the stone was rediscovered and soon afterwards it was built into the facade of the Cathedral. There it stayed until 1885 when it was once more moved, this time to a nearby museum.

The carvings on the Calendar Stone depict the Sun God, Tonatiuh, in the center, surrounded by the four suns of the past, representing previous destructions of the earth by jaguars, hurricanes, a volcanic rain of fire, and a flood. According to Aztec myth, the present sun or epoch is to end in a destruction by an earthquake and this symbol surrounds the image of the sun god in the center of the stone. There are also symbols representing divisions of the Aztec year including the twenty day signs of the Aztec month.

One modern note in the hall is a mounted specimen of a quetzal. This beautiful bird known to the Aztecs occupied an important role in their religion. Near the quetzal is a reproduction of one of the columns from the famous Temple of the Jaguars at Chitzen Itza. This Mayan carving represents the feathered serpent. It is shown with its rattles, its fangs like those of a rattlesnake and instead of scales has its body covered with feathers. The rattlesnake also figures in Aztec and Mexican lore. The legend of the original Aztec settlement of the capital site holds that the omen that determined the actions of the wanderers was an eagle which bore aloft in its talons a writhing rattlesnake. This event has been memorialized on Mexican stamps, money and on the seal of Mexico.

The Museum also has a fine cast of the round, flat, sacrificial stone believed to have been carved by the order of

Tizoc who ruled from 1482 to 1486. It was supposed to have been a memorial offering made at the time of the completion of the great temple to the Aztec God of War. The center was cut out to form a small bowl-like depression and it was in this bowl that the heart and blood of the sacrificial victim were placed. The Aztecs had other flat platform-like stones on which gladiatorial contests were held between selected prisoners of war and trained soldiers of the conquerors. The weapons usually consisted of flat, paddle-like wooden swords whose edges were set with razor-sharp flakes of obsidian, and short obsidian daggers. Some reconstructions of these weapons are on exhibit in the hall. The winner of such a contest, or rather of a series of such contests might be given his freedom if he were a prisoner of war. On the other hand a man who fought bravely might be granted the special and dubious honor of being offered up as a sacrificial victim.

The carvings and temples shown in the hall are impressive when one considers they are the product of a relatively primitive people, but they become even more a source of wonder when one realizes that the craftsmen who fashioned them worked without iron tools and built many of their temples without ever having learned the secret of the true arch. Most of the objects which are original pieces in the collection are of pottery, stone, bone, shell and metal. This is because these materials have withstood the ravages of time. Thus, the wooden Aztec drums and the obsidian mirror with its frame of carved wood represent great treasures and are part of the mere handful of Aztec woodcarver's art which has survived the Conquest. The Museum exhibit shows a number of smaller carved pieces including a fine figure of the ancient **Aztec Corn Goddess** from Ixtapalpa, Mexico. This piece is carved in black basalt and is an outstanding example of Aztec art. Numerous pottery pieces, many of them in the satirical and humorous style which characterize the ceramic art of western Mexico can be seen in the Hall of Mexican and Central American Archaeology.

One of the most recent and dramatic acquisitions to the collection is a cast of one of the giant Olmec style heads carved in diorite found in San Lorenzo—Tenochtitlan, Vera Cruz. This great head, not unlike a modern helmeted hero of the gridiron, has been placed in a central position in the hall facing its entrance. It stands about 9 feet high, is some 6 feet in width and in the original weighed 15 tons. Even the cast built as lightly as possible weighs several tons. It serves as a striking example of the skill of the ancient American artisans whose magnificent culture, now gone, can only be seen in the ruined remains of its larger cities and in the exhibits like that in the American Museum of Natural History. It is believed that the Olmec style head was carved in the first centuries of the Christian Era.

The Hall of Indians of South America, also on the second floor contains Indian exhibits from most of the countries of South America.

An exhibit in the Hall of South American Indians that illustrates a situation in which the patient method of archaeology was rewarded is the display which might be entitled the "Tax Collector." In 1896 the Museum received a small pre-Columbian silver figure, with no archaeological data save that it came from Peru. The Museum archaeologists who examined it could not imagine what kind of person the ancient silversmith had portrayed. It was wearing a curious headdress of a type not known and held a star-shaped stone headed club in one hand and a strange pinched cup in the other. The specimen was put away by the patient archaeologist to await the receipt of further data.

Thirty-seven years later, an elaborate feather headdress was found in a Peruvian mummy bundle, its use and age were unknown and it, too, was filed away in the reference collection.

Then, one day when the Museum archaeologist Junius Bird happened to be taking an inventory of this old material he was struck by the similarity of the actual headdress and that worn by the silver figure. Further examination of the stored collection turned up three more items; a war club, a cup and another portion of the headdress. The club was unmistakably the kind that the silver effigy carried and the cup was identical with that held in the hand of the metal figure. The complete headdress resembled the headdress and feathered robe worn by the tiny image.

Thus what was first seen in miniature in the silver figure later turned up in actual specimens. The mystery of the identity of the figure still remains unsolved but archaeologists think it might have represented a tax collector, his outstretched hand bears the cup for contributions and the club is held ready as though to help speed up the transaction. The figure has been humorously dubbed the "Tax Collector" and perhaps time and future discoveries may confirm the name or prove that it is wrong.

The ancient people of Mexico and Central America had a written language but the ancient South American Indians did not. These people who reached their cultural peak in the Incas of Peru were clever artisans, built large cities, laid out an elaborate system of roads and displayed considerable agricultural skill. They raised potatoes, oca, beans, maize, cocoa and cotton. They had succeeded in domesticating the llama and excelled in making pottery, producing fine metal work of copper, gold and silver and had mastered the art of fine weaving and embroidery. Their culture, like that of the Aztecs and Mayas, fell before the onslaughts of the Spanish conquerors and much of what we know of it has of necessity been deduced from evidences found in their ancient city sites and their mummies, wrapped in many layers of finely woven cloth made of cotton, wool and feathers.

The ancient Peruvians raised cotton, they obtained wool from the alpaca, and a domesticated variety of the wild vicuna. Bird feathers were often tied to a cloth backing to make capes and robes. They were the outstanding weavers in the New World and among the finest the world has ever known. Their woven textiles are highly ornamented with rich geometric patterns which represented the four motifs which continually occur in Peruvian decoration, the human figure, the bird, the fish, and the puma. The Museum has a display of some of the better examples of this extraordinary textile work and also possesses a number of mummies still wrapped in their heavy layers of cloth. Apparently the people of early Peru believed in a hereafter because food, utensils and weapons were placed in the bundles. In the mummy bundles of females, quite often a work basket was wrapped up and thus we have had a chance to study their looms, their threads, spindles and spindle whorls and all the paraphernalia they used in producing their cloth.

A large exhibit of Peruvian pottery is displayed in this hall including several good examples of the unusual whistling jars. These double-chambered clay pots or jars were filled with water. When the jar was tipped the water rushed from one chamber to another thus setting up a current of air which activated the built-in whistle.

Although the Incas may not have been as artistically inclined as the Maya, they had a political organization which far outranked that of any other of the governmental systems devised by the Indian. To bind their great empire together they built a great highway system. They constructed irrigation canals and ridged the steep slopes of their mountains with stone-walled terraces to increase the areas available to tillage. Their civilization was centered about large cities although the coastal communities were engaged in fishing and there were some hunting groups inland. Their masonry was amazingly good. They built large walls and structures and the neatness with which the joints in their stone structures fit together is a source of wonder. They did not use mortar and so well cut are the stones in their structures that even today a knife blade cannot be inserted between them. Like the Aztecs and the Mayas they had to do their stone work almost entirely with stone tools.

The Incas had learned to alloy tin from Bolivia with copper to make bronze and this superior metal was useful in certain crafts. They also worked gold and silver but most of their gold suffered the same fate as that of the Central American gold, being melted down into bullion for shipment across the Atlantic in the treasure galleons. The Peruvian mummy bundles, however, often contain gold and silver objects and thus furnish additional evidence of Peruvian metal craft.

A number of mummy bundles have revealed the fact that the ancient Peruvians practiced trepanation, a relatively advanced form of surgery used to relieve pressure on the brain due to skull injuries.

The dry climate along the coast of Peru has made it possible for these mummies to be fairly well preserved. The Museum's famous "Copper Man" was not found in a mummy bundle but was naturally preserved. It was found in a copper mine in Chuquicamata, Chile and appear to have been the body of an Indian miner who was killed by a cave-in while digging out the copper ore used by the Indians in their metal work. The tissues were impregnated with copper salts which have served to preserve it until the present day. With him were found the crude stone tools he used in his work as well as a sack he used to drag out the bits of ore he loosened from the sides of the tunnel in which he worked. He lived long before Columbus discovered America and was undoubtedly a slave.

In addition to the collection of material typical of the vast Inca Empire the Hall of South American Indians contains exhibits on some of the living tribes of Indians in Brazil, Ecuador, Chile, Peru and Patagonia. The most primitive of the Amerinds are probably the Tierra del Fueguans, who, living almost at the very end of South America withstood the ravages of a harsh and demanding environment with the crudest of shelters and weapons. A miniature group shows the Chama Indians in a tropical rain forest in northeastern Peru, along with their thatched dwellings, which are nothing more than wall-less roofs, their mode of cooking, their hammocks for sleeping and their dug-out canoes. A nearby case shows examples of the handicraft of the modern South American Indians—looms, pottery, wood carving, paddles and blowguns. One is struck by the simplicity and drabness of their decoration as compared with the more colorful ancient examples of handicraft.

A special case contains an exhibit of the shrunken heads of the Jivaro Indians of Ecuador, valued by the Indians as trophies of war. Some of the heads apparently once belonged to men of rank as they are elaborately decorated with colored feathers and iridescent beetle wings. The Jivaros prepare them by removing the bone and filling the head with hot sand, which, along with the subsequent smoking of the heads over a fire, causes them to shrink until some of them are about the size of a clenched fist. The Jivaros used this process to shrink not only human heads, but those of animals also, as several shrunken sloth heads on exhibition indicate.

The North American Indian exhibits in the Museum are housed in five halls, devoted to the Indians of five major areas. The smallest of the halls but not necessarily the most unimportant is the Eskimo Hall. Here tools and weapons of bone, ivory, wood and skin and samples of the fur clothing and ceremonial equipment serve to reveal the material culture of the northernmost of the Amerinds.

All of the Eskimos now alive could be seated in the Yankee Stadium and about one-half of the seats would be vacant. These thirty thousand or so inhabit a vast and barren territory in Alaska, Canada and Greenland, most of which is north of the Arctic Circle. Life is hazardous when lived under the conditions the Eskimos face. Some years ago, Robert Flaherty made his now classic documentary film, "Nanook of the North." Nanook, the Eskimo hunter who starred in the picture, died within one year from the onslaughts of the rigorous environment he was shown combatting successfully in the motion picture. For the most part the Eskimos live along the seacoasts of Greenland, Labrador, across North America to Bering Strait and over to Northeast Siberia. The sea furnishes fish, seal and walrus, while on the land musk-oxen and caribou supply skin and meat. Most Eskimo are nomadic, travelling constantly in search of food. Since they are without agriculture they must be hunters and in a region where game is not too plentiful and where some game migrates, the hunter must seek out his quarry.

The Eskimos are thought to be the most recently arrived of the Mongoloids who populated America. There is no proof that they were forced to stay in such a rugged climate and it may be that they were already pretty well adjusted to life in the north and after coming to the New World continued to perfect their adjustment to living in a cold area. In any event they are good evidence of man's ability to become established wherever enough food to support life exists. They are an ingenious people and have made good use of the rather limited resources the area makes available to them. Their invention of skin boats, of snow-goggles, of dog sledges, of harpoons, of a snow hut for winter use and of the seal-oil lamp, are all indicative of their ability to master their environment. The numerous small ivory and bone carvings of the animals of the region reveal a creative love of beauty. Life is at best hard in the far north. The Eskimo hunter needs to expend most of his energies in the struggle to stay alive yet the Museum's collection of these carvings reveal a close observation of nature coupled with considerable skill in handling carving tools. Many items in the Eskimo collection were brought to the Museum by the expeditions of Robert

E. Peary, Donald MacMillan and Vilhjalmur Stefansson.

Though the Eskimos have been forced to get along with little or no wood, their neighbors to the south, the North Pacific Coast Indians have an ample supply which has enabled them to develop into the most skillful wood workers on the American continent. A quick glance at the Jesup Hall of the Indians of the North Pacific Coast shows an abundance of items made of wood from huge totem poles to a still larger sea-going war canoe hollowed out of a single cedar log, from hundreds of masks carved of wood to ornamental boxes (original cedar chests) and finely carved trays. The far-reaching Jesup expeditions brought back this fine collection and the walls of the hall are embellished by a series of murals by Will S. Taylor which illustrate the industries and the religious and social activities and which also give a good picture of the heavily forested, fog and rain drenched environment of these capable and artistic people.

The centerpiece of the North Pacific Coast Indian Hall is a large Haida canoe. This seaworthy craft, about 64 feet long and 8 feet wide, could hold fifty men. The Haida have been known to make voyages of over five hundred miles in such canoes. Figures have been placed in the canoe, some shown dipping their paddles into the water near the stern and others standing in the prow are shown holding the boat off shore with long poles. The canoe is supposed to be ready to be beached and in the stern are a group of Indian medicine-men and other dignitaries dressed in elaborate masks and headdresses. The hold is loaded with boxes and it is evident that this is not a war party. The group is supposed to represent a party of Indians on their way to celebrate a potlatch.

NATIVE AMERICANS

One of Will Taylor's murals shows a potlatch, a social ceremony which brought high prestige to its giver. The word "potlatch" means "to give." To hold his place in the Indian aristocracy or perhaps, to gain a place in a higher social group, the Indian would give a potlatch.

Blessed with an abundance of nature's riches, these Indians nevertheless placed great stress upon the possession of property. The highest group in the community were not the priests or the warriors, but the men of wealth. The wealthy Indians of the Northwest formed a sort of aristocracy. They held captives and kept the descendents of their captives in slavery, granting them no community rights. The common people, midway between the slaves and the aristocracy, could climb the social ladder by the accumulation of wealth.

To the wealthy Indians of this group honors and names were of paramount importance next to wealth. Thus the custom of displaying one's totem in front of his house developed. The totem was in a sense a pictorial testimony of the individual's family history and through it the owner publicized his claim to fame. The Museum exhibit shows a collection of totem poles each carved with the animal figures and semi-spirit ancestors of the owner, or embellished with symbols which signify some important and praiseworthy aspect of the owner's background.

Two of the tribes represented in the collections, the Shushwap and the Thompson lived in the interior of British Columbia. All the rest are coastal peoples and the exhibit presents them in the order in which one would encounter them in travelling from south to north along the coast of Washington, British Columbia and Alaska. The tribes thus depicted are the Bella Coola, Tsimshian, Haida, the Nootka, Kwakiutl and the Tlingit.

These Indians worked in wood, and, having no metal used only stone and bone mallets, scrapers, awls, axes and wedges. They had nothing resembling a saw and they did not possess nails. They felled trees with their stone axes and burned the fallen tree trunk into log sections by building fires under it. They scraped off the bark and then by rows of skillfully driven wedges they split the logs into boards which they adzed smooth and then finished with sharkskin which they used as we use sandpaper. In place of metal nails they drilled holes in the planks and lashed the pieces of wood together with spruce roots.

The large wooden houses built by some of them were conveniently placed along the shores of the sea or on the banks of rivers. Some of these houses were so large that their construction had to be a community affair. Large timbers and beams formed the frame and the walls, and roofs were made of planks laid in an overlap so as to shed the rain. When the Indians moved off to their summer hunting and fishing grounds they simply untied the planks and carried them by boat to their other camps. These planks were of great value since they were so difficult to make with the crude tools the Indians used. By removing them the Indian needed only one set for both his houses. The heavy house timbers were left in place, their very weight and cumbersomeness insuring them against being tampered with by any chance marauders of the empty village.

The Museum exhibit shows in carefully made models the nature of these plank houses and also shows the way trees were felled and the planks split out of the logs. Many fine examples of boxes are also shown, some of which were used for cooking. The boxes were waterproofed, filled with water and then hot stones were dropped into them which caused the water to boil and thus the food was cooked.

The Indians often used the inner bark of the cedar for making clothing. They stripped the bark from the tree, then shredded it into long fibers and wove these fibers into cloth. They also made handsome blankets from the wool of the Rocky Mountain goat and made their horn spoons from the horns of both the Rocky Mountain goat and the bighorn sheep. They were ingenious hunters and trappers and a display of model fish traps and animal snares show their skill.

If the Indians of the North Pacific Coast are to be typified by their skill in wood-working, then the Indians who lived on the treeless plains of the great west and whose story is told in the Plains Indian Hall are to be characterized by their clever use of skin and hide. These people were nomads, forced to move about because their principal source of food and skin, the bison or buffalo, moved about in search of grazing land. Their life pattern was adjusted to their chief occupation, following the buffalo. Nomads have to travel light, they cannot afford to accumulate bulky possessions. Before the introduction of the modern horse into the New World by the Spaniards, the only beast of burden these Indians possessed were dogs. Another handicap which they shared with all the other Indians in the Americas was the lack of a wheel. Since they did not have the invention which might have made it possible for them to build carts or wagons, they devised a kind of drag made of two long poles crossed over and lashed together near one end. The narrow end was placed over a dog's back and tied in place with a sort of harness and a small carrying platform of sticks and thongs was lashed across the poles at a place where the long poles spread out. The ends of the poles dragged along on the ground. The same type of drag or *travois* on a larger scale was used with the horse when the Indians finally learned the value of the white man's principal beast of burden.

More important than the *travois* was their invention of the tipi, a tent-like structure made of buffalo hides stretched over long poles, an ideal dwelling for a nomadic people. It could be easily taken down, carried to a new location and in a matter of minutes, set up again. The Hall of the Plains Indians is dominated by a large Blackfoot Tipi. Hundreds of items typical of the culture of the Plains-Cree, Dakota, Crow, Blackfoot, Gros Ventre, Arapaho, Cheyenne, Mandan, Hidatsa, Omaha, Kansa, Iowa and Pawnee Indians are exhibited in the hall, but the entire collection does not occupy nearly as much space as that of the Indians who lived in large, relatively permanent village sites. Like the Eskimo, these people were forced by the nature of their life as nomads to limit their possessions to only the most essential items.

When the westward movement of the settlers began and when the white hunters made great inroads upon the vast herds of buffalo, these Indians were forced to leave their hunting ranges and gradually their numbers decreased. They were among the last of the Indians to withstand the invasion of the white man.

One of the exhibits shows some of the Ghost Dance shirts. These flimsy cloth shirts were worn during the Ghost Dances which accompanied one of the last great attempts to oust the invaders. The ghost of the buffalo that had been killed by the white hunters was supposed to return; the Indians were supposed to be restored to their former lot in life and the white man was to be driven away. So enthusiastic did the aroused Indians become that some of them wore these flimsy shirts into battle under the belief that they were bullet proof.

The Sun Dance was one of the most important ceremonies of the Plains Indians. The Museum exhibit contains a model of an Arapaho Sun Dance. In this miniature group can be seen the ceremonial lodge, a circular ring of poles and thatch covered with a roof of poles all converging on a center pole. A group of interested spectators watch a score of Sun Dancers who have whistles in their mouths and are keeping their eyes focused on the sun as they perform their shuffling dance steps. Off to one side four or five drummers are beating out the rhythm. Offerings to the Sun are hung on the center pole of the lodge and on the left side a kind of altar is erected. Sun Dances were held in mid-summer and were occasioned by an individual taking a vow. Sometimes the vow was one of vengeance, sometimes to placate the Sun God. Certain features of the dance were designed to test the dancer's endurance. One was to gaze into the sun, another to fasten skewers of bone under the skin of the chest, attaching a buffalo's skull to the skewer by means of a thong, and then, while dancing, attempting to pull the skewer out of the skin. When the United States Government put the Plains Indians on reservations it forbade the inclusion of these features of self-torture in the Sun Dance.

The stereotype Indian headdress is typical of the Plains Indians. No other Indians had such elaborate and long feathered headdresses. The important feathers in any brave's headdress had to be won by special deeds of bravery in battle. The more of these feathers a warrior accumulated, the higher his rank in his group. When the Indians learned to ride the horse and for some two hundred years enjoyed its use before the ultimate invasion of their territory by hunters and pioneers, horse stealing became a principal feature of their warfare with neighboring tribes. An Indian adept at this practice might accumulate enough coups to rise rapidly in rank.

The importance of the buffalo to the existence of the Plains Indians cannot be overestimated. The great herd of buffalo that roamed the plains in the days before the coming of the white man have been estimated at over 65,000,000 head. The Indians would try to keep in contact with the herds during their summer and winter migrations. The main supply of meat and pelts was obtained during the great tribal ceremonial summer hunt when the animals were fat and their hair was thin. The flesh of the buffalo was then in the best condition for food and the thin hair made it easier to dress the pelts on both sides. The Indians used buffalo hides for making clothing, shields, packs, bags, ropes, snowshoes, tipis and skin boats. In order to preserve the meat it was cut into thin strips which were hung on frameworks of poles to dry in the sun. This sun-dried buffalo meat was the famous "Jerky" or "jerked meat" used by the pioneers. It was thus called because the thin strips of meat were practically jerked or torn from the carcass. Indians used the sinews of the animal for bow strings, thread and rope and the buffalo hair for weaving lariats, belts and personal ornaments. The horns were made into drinking vessels and occasionally spoons.

Buffalo were hunted in winter but the hunting parties were small and independent and were not subject to the many rules and regulations which involved the summer or ceremonial hunt. Pelts secured in the winter were used for bedding and heavy garments of extra warmth because of the thicker coat of winter hair the buffalo then possessed.

Some of the Plains Indians supposed the buffalo to be the teachers of the medicine men instructing these men where to find healing plants and herbs. The buffalo played a part in many of the Indian ceremonies and was always spoken of in a most respectful manner. It became a symbol of long life and plenty to the Plains people, and when the buffalo were practically exterminated in the last quarter of the 19th century the Plains Indians suffered a severe blow.

A long time ago a nomadic group of Indians settled into the southwestern part of what is now the United States. Little is known about these early forerunners of the present Southwest tribes except that they were nomadic hunters without an agriculture. Their descendents, however, became farmers and raised maize, squash and beans as their principal crops. They developed into skillful weavers of baskets and they have been called "The Basket Makers" by the archaeologists. The Basket Makers were replaced by another group of Indians who were to become the ancestors of the builders of the pueblos, the houses made of adobe brick.

The Hall of the Indians of the Southwest deals with four major groups of Indians from New Mexico, Arizona, western Texas and southern Colorado. It presents material from both the prehistoric sites of this area as well as from the living tribes. The dryness of the climate of this desert country has preserved the ruins of some very ancient village sites. Before the building of pueblos some of the sedentary Indians of the area lived in cliff dwellings. The Museum exhibit contains models of the early cliff dwellings as well as the more outstanding examples of more recent pueblos.

Not all of the Southwest tribes were village dwellers. The Apache, Navajo, Pima, Papago and Havasupai were nomads. The Museum exhibit contains a considerable amount of Navajo and Apache material. The Apache live in grass thatched houses or in the open under the shade of flat-topped, open-sided shelters. They have been called the greatest natural horsemen in the world. It was the Apache, under Geronimo, who raided the settlements of southern Arizona and northern Mexico and eluded the United States troops for many years. They practise some agriculture but this is scant compared to the more extensive agriculture of the pueblo Indians. In many respects their culture is the poorest of the Southwest groups.

The Navajo on the other hand raise corn and have large flocks of sheep whose wool they use to weave their colorful blankets and whose flesh they eat.

Life-sized groups show the Apache and the Navajo and the Hopi Indians. The Navajo are shown with their sheep, their Spanish-influenced clothing and their silver concha belts and squash blossom necklaces. The scene shows some Navajo medicine men performing one of their healing and religious rites on a man and a boy. Part of the group shows the interior of the medicine lodge or hogan with a medicine man preparing a sand-painting. These "paintings" are made by sprinkling colored sand on a base of gray sand and drawing thereby the figures of gods and other religious and sacred symbols. They were considered very powerful medicine by the Navajo. Nearby cases show many good examples of Navajo silver craft and of the brightly colored Navajo blankets.

Most of the Indians of the Southwest today have a hard time eking out an existence. Prolonged drought and the meager vegetation make it difficult to raise sheep and the little farming that is done is just about enough to keep the Indians from starvation. Museums try to encourage the handicraft of the natives and the Indians make some money by selling their silver work, blankets and pottery. Some entrepreneurs have engaged the Indians in mass production of these items but this often results in a high loss of quality. Good Southwest pottery, jewelry and blankets are not cheap. They are carefully made, often are works of art that have taken their creators a considerable amount of time to complete. The University of New Mexico at Sante Fe has done much to encourage the perpetuation of these arts and crafts. Colorful ceremonies attract many tourists and some extra income accrues to the Indians at these gatherings.

One of the large groups in the hall shows a scene on a Hopi pueblo. Here we get a glimpse of life in an Indian "apartment house." Built in tiers of adobe, some of these pueblos rise five stories above the ground. Quite often they were set on top of a mesa, one of the flat-topped, table-like prominences so common in this area. Originally these tiered houses had no entrances or openings on the ground floor. Access was by ladder which enabled the Indians to defend their villages more easily. Today, with the need for such protection eliminated, the modern pueblos can be entered from the ground level. Near the pueblo there are usually one or more underground chambers or kivas. These are used by the men for their secret ceremonies.

Coming under Spanish influence at an early date, the Indians of the Southwest present today a culture which is part original Indian, part Spanish and part an accumulation of more recent contact with modern civilization. For example, there is a young matron of Indian descent who lives in a modern apartment in the town of Sante Fe. Once a week she gathers up her laundry and travels a short distance to the pueblo of Tesuque. Her Indian aunt lives there, and since Tesuque has been recently equipped with electricity, the elderly Indian woman has an automatic washing machine which she allows her more modern niece to use. The washing machine stands in the single room which is both living and sleeping quarters for the family.

The rooms in an adobe pueblo are quite comfortable. The thick mud-brick walls insulate the rooms, keeping them cool in summer and warm in winter.

Small dioramas show the Snake Dance and other ceremonies of the Southwest Indians and there is an interesting exhibit which shows how some of the prehistoric pueblos have been dated by using a method which is based upon a study of tree-rings. This is done by studying cross-sections of the roof poles in the old pueblos. Two great collections, one of baskets and the other of pottery show how adept the Southwest Indians were in these crafts.

The Hall of the Indians of the Eastern Woodlands deals with the aborigines of northeastern United States and Canada but also includes the Southeastern and Mackenzie culture areas. Before Colonial days this whole area was forested from the Atlantic coast practically to the Mississippi River. Assuming that the fundamental ideas of North American Indian culture travelled northward out of Mexico, and that each tribe did not pass on to its neighbors to the north all it had learned, the northernmost tribes would then possess very few of the elements of the higher Indian civilization in Mexico. Thus the northernmost tribes of the Woodlands area have much less advanced ways of living than those of the Southeastern tribes.

The Eastern Woodlands Indian Hall begins with the tribes closest to New York and the subsequent exhibits deal with tribes ranging westward and northward. There is a special exhibit dealing with the Indians of Manhattan Island. For some time after this strip of land was bought and colonized by the Dutch, a few Indians remained on the scene, but they were soon to move out as the settlement grew. Most of the items in this special exhibit were dug up at old Indian village sites of which there were a dozen or more on Manhattan.

The only site on the island still in its original condition is at the northern tip of the island in Inwood Hill Park. Here, shaded by giant tulip trees that are hundreds of years old, are some caves made by the fallen slabs of rock from the ridge above. These rock shelters were used by the Indians and the dirt on the shelter floors has been thoroughly screened and sieved. Many items of Indian culture were found in the area. Nearby there still can be seen a layer of shells which represent the shell heaps built up by the Indians who ate the shellfish and used parts of the shells for their wampum beads. There is a reproduction of these rock shelters in the Eastern Woodlands Indians Hall along with some of the artifacts found at the site.

From the brief accounts given of the Manhattan Indians by the early colonists and from the artifacts dug up at the old village sites the Museum has attempted to reconstruct the culture of these early New Yorkers. Once while excavating for a building site in lower New York, part of an Indian dugout canoe was found. This is also on exhibit. The characteristic dwelling of these forest Indians was a conical structure covered with slabs of the bark of birches or other trees. The Manhattan Indians used a dome-shaped, instead of a conical wigwam.

The Eastern Woodlands Hall contains a fine series of dioramas which depict various Indian activities. Two of these show the way the Indians planted maize, harvested it, made it into meal and baked it into flat cakes. Other groups show the many uses of birchbark, the typical Indian farm, the processing of deer skins, the making of maple syrup and other features of Indian life. A full-sized real Ojibway wigwam is also on exhibit. This is the conical-shaped type and is covered with birchbark.

The Iroquois exhibit deals with the material culture of these remarkable people. An exception to the rule that as one progresses northward one finds more primitive groups, these people developed the concept of the confederacy to a high point. They expanded their system into what was known as the Iroquois League. Certain member tribes were assigned particular and specific responsibilities which concerned the welfare of the Six Nations. For example, the Mohawks were charged with the responsibility of keeping the peace. They were the "dog-soldiers" of the League. The other tribes which made up the League included the Seneca, Oneida, Onondaga, Cayuga and later the Tuscarora.

The forest Indians of the East were primarily hunters and fishermen. Wherever climate permitted they practised an agriculture in which women did most of the work. Wild rice and maple sugar were staples, but corn, beans, squash and tobacco were also raised. They used wood for canoes, mortars in which they ground the maize, spoons, bowls, dishes and other utensils. Their pottery was poor compared with that of the Indians of the Southwest. Their principal weapons were the club and the bow and arrow. The famed "tomahawk" was nothing but the whiteman's axe which the Indians soon adopted for their own use. Later they obtained guns from the settlers and more and more they began to include in their material culture items which they obtained by trade or raid from the settlers. They were doubly affected by the coming of the whites, losing their land as the whites increased and losing certain aspects of their culture by living in close proximity to the white men and working for them as hunters and trappers or as military allies.

Though the Indians learned much from the settlers they in turn were able to introduce the settlers to many of their own agricultural products. A case in the hall shows samples of agricultural products which the Indians had developed and which are even today an important part of our diet and materia medica. Perhaps the most important of these was Indian corn or maize but other important Indian crops were beans, squash, tomatoes, tobacco, chocolate, coca (cocaine), peanuts, and cotton.

The exhibit contains many fine examples of beautiful Indian beadwork and needlecraft. They adopted their designs from nature, and the moccasins, leggings, and shirts are embellished with floral designs. When the finer, stronger steel needles brought by the white men replaced their crude bone needles they produced some excellent needlework. They did some weaving but more often relied upon trade cloth which they then proceeded to decorate Indian fashion.

The Eastern Woodland Indian was the model for the traditional noble savage of song and story. True, he did not have the elaborate feathered headdress with which he has been credited. That belonged to his Plains brothers. He may not have been as advanced as his southern and western cousins but he had a culture sufficient to meet the needs of his environment. Some of his descendents have been absorbed into our culture, others are still "long house" or orthodox whether they live on the reservation or off it. Like many of the excellent Mohawk ironworkers who have formed a little colony in the Bushwick section of Brooklyn, New York, on workdays they may be New Yorkers, but when they are on their own time, they are Mohawk Indians who still retain their old tribal customs and who frequently go back "up-state" to visit the old folks on the reservation and to take part in the ceremonies that were old when Henry Hudson was but a lad.

CHAPTER EIGHT

The Twain Shall Meet

THE EXTENT to which a single individual may influence the destiny of an institution as large and complex as the American Museum of Natural History is amazing. Morris K. Jesup served as a member of the original board of trustees of the Museum, was its third President, and richly endowed the institution in his will. During his lifetime he endowed the series of anthropological expeditions that went into Alaska, British Columbia and Siberia. The extensive collection of ethnic materials from the relatively little known natives of Northeastern Siberia—the Koryak, the Kamchadal, the Chukchee, the Yukaghir, the Tungus, the Yakut, and the Churantzep—now on display in the Asiatic Natives Hall was acquired by Waldemar Jochelson and Waldemar Bogoras on the Jesup North Pacific Expedition. In addition to some 8,000 objects which illustrate the life, customs, belief and culture of the tribes in this area, Jochelson and Bogoras brought back plaster casts of faces, skulls and archaeological specimens from graves and abandoned village sites as well as field journals filled with hundreds of the folk tales and traditions of the people. They also obtained phonograph recordings of native dialect, songs and chants and hundreds of photographs.

The coastal natives of northeastern Siberia, like the North American Eskimo, depend on fish and other forms of sea life for their existence, while those farther inland depend upon the reindeer. Both the Maritime Koryaks and the Reindeer Koryaks dressed in reindeer skin. The Chukchee and the Koryak have both coastal and inland groups, in the case of the Koryak about equal in size, each numbering about 3,500 people. The Reindeer Branch of the Koryak keep herds of reindeer and live almost entirely on the meat of these animals. They live in movable tents while their maritime brothers live in houses built partly underground.

The Museum has a model of one of these houses which demonstrates how well designed it is to protect its inhabitants from the severe winters of this region. It is built of logs, poles and earth. The lower portion is a truncated cone, not unlike the lower part of a Plains Indian tipi, but the sides have been allowed to extend upward and outward so that they eventually form a sort of inverted cone on top of the cone-shaped main part of the house. The main part of the structure is almost half underground and earth is heaped on its sides to keep out the wind and cold. The inverted upper cone acts both as a smoke hole, and as a means of keeping the roof entrance free from drifting snow.

The collection contains a fine display of the excellent carvings in wood, antler, and ivory, produced by the Koryak who also are skilled in making tastefully ornamented fur rugs. Until fairly recent times, when they came into contact with other cultures they had only stone tools. In recent times, however, they have become fairly skilled in the blacksmith's art. A life-size figure on exhibit shows a Koryak man dressed in the unusual but effective armor made of plates of iron lashed with leather thongs on a kind of skin coat. A folding skin shield is supported by thongs from the neck, shoulders and arms, and protects him from arrows which may come from behind or from either side of his head and shoulders.

The Yakut live in north central Siberia and depend upon the horse both as a beast of burden and as a source of food. Their staple liquid food is mare's milk which they make into kumiss by allowing it to ferment and churning it in large skin churns. The exhibit shows examples of their churns as well as an assortment of handsomely decorated vessels of poplar wood from which the kumiss is drunk. Some of these drinking vessels are large enough to hold several gallons and must have been used as communal cups.

The rest of the exhibits in the Asiatic Natives Hall contain material from the Andaman Islands in the Bay of Bengal, from Burma, Korea, Japan, Tibet and China. None of these collections are extensive and most of the materials were collected half a century ago. However, the Chinese collection contains some fine ancient bronzes which may be 2,000 years old and an interesting display of Chinese ceramics. The rest of the Chinese material consists of items which were concomitants of everyday life in China a half century ago. The bulk of this collection was obtained by Dr. Berthold Laufer during a period of three years in China. Many changes have taken place in China since then, and the collection now has a special significance in that it represents a phase of Chinese culture that has passed away.

The small but colorful Japanese collection is interesting as an example of a civilization that stems from the Chinese. Long ago the Japanese came to their islands from the Asiatic mainland. They there developed a form of life which appears to have been based on Chinese customs and ideas but which soon came to have a distinctive quality of its own. The objects shown in the collection display the characteristic precision and fine workmanship of the Japanese. Things have changed greatly in Japan too since the collection was obtained, and it, like the Chinese collection, has now an added significance.

The Museum also has an exhibit of artifacts of the Ainu —the hairy aborigines who inhabited the northern islands of the Japanese group before the race later known as Japanese invaded the islands. The Ainu material indicates a very primitive culture which had its beginnings at least as early as the New Stone Age. Relics found in shell deposits in the most southern parts of Japan would make it appear that they once ranged over a much larger area than they now occupy. Then as now they were simple hunters and fishermen. They had no agriculture and the only animal they had domesticated was the dog. They had bows and arrows, made some pottery, and used a ground stone axe as their principle tool.

The Museum is fortunate in having a small but very representative Tibetan collection consisting mainly of objects used in the ceremonies of Buddhism, whose lamas or monks are Tibet's ruling class. Most of the Tibetan men are monks and the support of the many monasteries absorbs a large part of Tibet's wealth and man-power. The monasteries are large, handsome stone buildings usually set on a mountain top. Some of them hug the steep mountain sides so closely that they seem parts of the rocks themselves. Their religious art has been developed to a high state.

The Museum exhibit contains many beautifully painted religious scrolls and banners depicting one or another of the hundreds of dieties of Tibetan Buddhism. There are also many bronze statuettes of the Buddhist gods and goddesses as well as temple paraphernalia. Included among the latter are some drinking vessels made of human skulls, a small trumpet made of a human thigh bone and several large metal temple trumpets or shawms. These great metal horns sometimes are ten to twelve feet long. One end is rested on the ground while the horn is blown to call the people to religious ceremonies.

The Mongoloid Tibetans live in a mountainous land where some of the highest peaks in the world are to be found. Those who live on the wind-swept plateaus are nomadic while those living in the deep river valleys tend to congregate in settlements. The climate of the country is one of the harshest to be found anywhere in the world. The life of the common Tibetan is primitive and unenviable. They use the native yak as beast of burden, as riding animal, and as the chief source of meat supply, butter, cheese and milk. The hair of the yak is woven into a strong felt-like cloth.

The Museum exhibit also has life-sized figures of a lamist priest and a well-to-do Tibetan woman dressed in full attire. The man has in one hand a portable prayer wheel. This ingenious device consists of a little metal drum mounted on a handle on which it can spin around. Little weights are fastened to the drum by small metal chains and they serve to add impetus when the owner whirls it in his hand. Prayers written on paper are stuffed into the hollow drum and every time it revolves the prayers are supposed to be carried up to the Tibetan heaven. Thus a man may be engaged in other religious activities or in more mundane everyday duties and still be offering up a constant stream of prayers so long as he remembers to keep the drum spinning around.

The American Museum of Natural History is but a little more than a mile from Harlem, one of the largest communities of Negroid peoples in the world. Many of the Negroes in America are descended from West African stock and yet the majority of them have little or no knowledge of the country of their forebears. True, their music crossed the Atlantic with them in the terrible days of the slave trade but their gradual acclimation to the ways of life in the New World was accompanied by the loss of much of their cultural heritage. Preserved in the African Ethnology Hall on the third floor of the Museum is a comprehensive collection of the specimens representing the arts and crafts of the Negroid peoples of Africa.

Northern Africa has received many waves of white immigration. Carthaginians, Romans, Moors, and Arabs swept into the northern third of the "Dark Continent" and left their mark. The great triumphs of ancient Egypt brought into the Nile Region peoples from non-Negroid lands around the Mediterranean. However, since prehistoric times the home of the Negroes, the Bushmen and the Hottentots, has been in central and southern Africa, covering two-thirds of the continent. The Bushmen and the Hottentots are closely related and although Negroid in character, they are yellowish brown in color and have other physical differences which differentiate them from the true Negro. The Museum collection contains materials related to these three groups. The items in the collection are arranged according to the various cultural and racial groups which produced them.

One of the more unusual collections in the exhibit comes not from Africa but from South America. About the middle of the seventeenth century some Negro slaves managed to break away from their owners in Dutch Guiana. Hiding in the tropical forest they established a colony which still exists. Their speech and customs are very much like those of their West African motherland. They even have continued the manufacture of wooden articles in the same manner and decorated with the same type of raised, serpentine design that was typical of the group when it lived in Africa.

The culture groups in the hall demonstrate different economies. Some of the peoples represented are herdsmen, others hunters and still others are farmers. One factor which distinguishes the Negro culturally from the aborigines of America, Oceania and Australia is the technique of smelting and working iron. One is struck by the abundance of tools made of this metal. There are also beautiful examples of carving in wood and ivory. Living in a land of wild game, the Africans have always had a plentiful supply of skin to furnish material out of which to make clothing and other useful articles. The pottery on the whole is not exceptional and baskets woven of fiber and reeds are often used as some other peoples use pottery, that is, as utensils to hold food and other items.

One of the showpieces of the exhibit is the Lang Ivory

Collection. This rare and beautiful array of objects carved in elephant tusk ivory, the former treasure of an African king, represents what is probably the last coordinated output of a West African kingdom, that of the Mangbetu. It is boldly realistic and not so much indicative of the genius of an individual artist as an expression of a national art, mirroring the aesthetic influence of the whole tribe, an example of some of the best in the artistic expression of a proud people. Today, with the source of the ivory greatly diminished, this art has been almost entirely lost.

Many of the forms are elongated as a result of the shape of the elephant tusks. The gleaming white ivory lent itself to drawing so that the Mangbetu craftsman had opportunity to represent his skill in the graphic arts. Thus, many of the pieces may be said to be engravings rather than sculptures. The collection includes a variety of articles. There are ivory trumpets, some partly sheathed in leopard's skin, ivory handled sickle-shaped knives, daggers, drinking cups, hair pins and combs, and some fine brushes, used to brush away flies and other insects from the presence of the king. They are made of the hair from elephant tails and the handles are of carved ivory. Since there is but little hair on a single elephant's tail even the number of brushes in this collection indicates that the tribe was either skillful in the hunt or wealthy enough to buy without stint from other tribes.

Another rare and beautiful African art form is represented by the valuable Benin bronzes. These bronzes and brass castings, made much in the way in which similar items were produced in the Renaissance period in Europe, that is, by the lost wax method, come from only one locality in West Africa. It may be that the art was introduced by Europeans

THE PRIMEVAL FOREST. The Felix M. Warburg Memorial Hall, the first of a series of exhibitions intended to indicate the interrelationships of Man and Nature in certain significant localities throughout the world. The Warburg Hall tells the ecological story of the Pine Plains region in Dutchess County, New York. The Primeval Forest is part of a panorama in five sections which depicts the broad external picture of the forty-square mile area at five stages of its development. Here in the first stage is a heavily forested area inhabited only by wild creatures and Indians. We see an Indian village set in a small clearing where a crude kind of agriculture produces small crops which supplement the Indian diet of game and wild plants.

THE SETTLEMENT—1790. With the coming of the first white settlers the Pine Plains region experienced the first impact of a culture which greatly altered its condition. One of the first tasks of the pioneers was to clear away large sections of the forest in order to obtain wood for their homes and space for their fields. Though the agricultural implements and methods of the settlers were better than those of the Indians, nevertheless they were crude and wasteful compared with modern equipment and techniques. In 1790, preparing the land for the plow meant many hours of heavy labor in removing stumps of trees and boulders from the soil.

THE HIGH TIDE—1840. By 1840 most of the primeval forest of the Pine Plains area had been cleared and put to the plow. The area presented an appearance of prosperity; the returns from the freshly cleared virgin soil were abundant. Farming was still hard work but farm machinery had improved and horse-drawn cultivating and reaping machinery were making possible the raising and harvesting of bumper crops. The intensive farming that was carried out in this era was not accompanied by any efforts to retain the fertility of the soil. Cover crops or timber stands to prevent erosion of the precious topsoil were not provided.

THE EBB—1870. Just about the time that the American West was being opened to settlement, the Pine Plains area was starting to reap the harvest of years of wasteful and destructive agriculture. Farm after farm was abandoned as the fertility of the soil grew so poor that farming in the area was no longer profitable. Some farms in the valleys kept going but on the ridges and slopes erosion had taken its toll. Families packed their belongings in covered wagons and started on the long trail to the fertile prairie lands of the West. Man had finally impoverished the once fertile soil of the Pine Plains region.

TODAY—1951. Man has reclaimed part of the Pine Plains area by the practice of soil conservation and intelligent agriculture. Trees once again cover some of the hills thus preventing the destructive effects of water. Use of modern fertilizers, crop rotation and contour plowing have made farming once more possible. Modern motorized equipment make the farmer's job easier and his efforts more effective than in the days of ox-drawn and horse-drawn equipment. The Warburg Hall contains many other exhibits which develop the story of soil-plant—animal interrelationships in this rural area. It is the purpose of the proposed additional Man and Nature exhibits to show the interrelationships of Man and the animate and inanimate aspects of his environment in other parts of the world in order that people may learn to make a successful adjustment to their natural environment.

Above: Roy Chapman Andrews (left) and Walter Granger (right) uncover the fossil bones of a giant animal of the past in the Gobi Desert of Mongolia.

Below: James L. Clark goes exploring via "Yak-back" in the high mountain country of Mongolia.

Above: Museum trustee Michael Lerner looks on as his catch, a large squid, is hauled on board.

Below: Dr. Robert Cushman Murphy (right), Chairman of the American Museum's Bird Department helps to display a Cahow. This bird, thought to be extinct was recently rediscovered by Dr. Murphy in Bermuda.

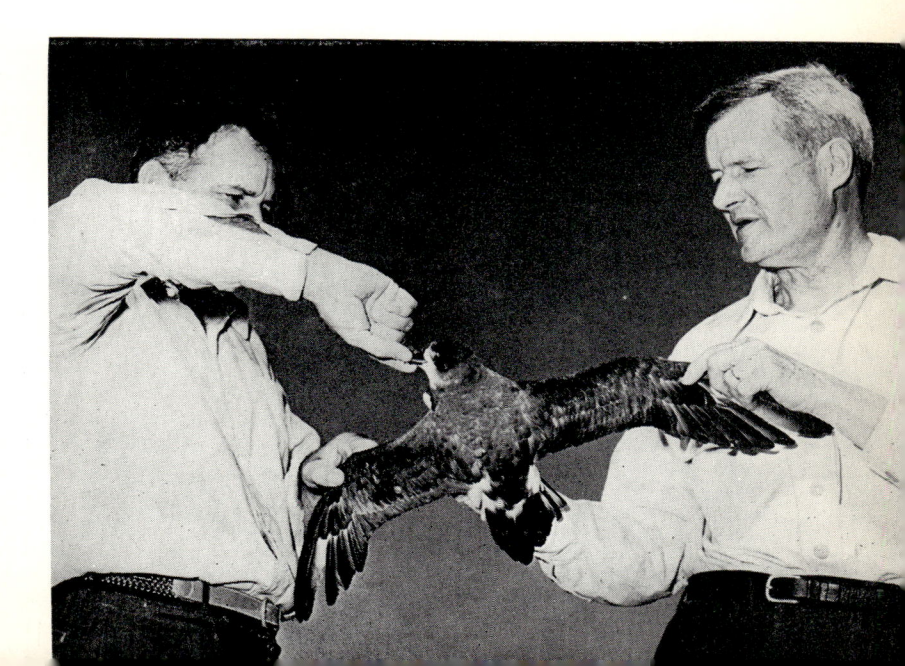

CARL AKELEY AT WORK. The late Carl Akeley did more to advance the quality of museum taxidermy than any other single individual. Akeley's talents as a sculptor, his keen interest in wild life and his inventive ability led to the development of the "Akeley" method of taxidermy. This method has resulted in the very life-like mounted specimens in the American Museum's beautiful habitat groups. Akeley African Hall at The American Museum of Natural History is named in his honor. Akeley explored much of Africa and his concern over the vanishing wild life of that continent led to his planning a great African Hall and also stimulated the Belgian government to set aside a large tract in the Belgian Congo, the *Parc National Albert,* as a sanctuary for the wild life of that region.

Robert Rockwell models a life-sized clay figure of the giant Alaska Brown Bear. Modeling the animal figure in clay is one of the steps in the "Akeley" method. A plaster-of-paris mold is then made of this clay figure. Inside the plaster mold a manikin of wire mesh and *papier-mache* is then made and the actual prepared skin of the animal is then draped over the manikin. The finished mount of this great bear is on display in the Hall of North American Mammals.

Above: Preparator George Adams and some of the birds he mounted for the colorful Tropical Bird Group in the Sanford Hall of the Biology of Birds.

Below: Paleontologist Barnum Brown (right) and preparator Otto Falkenbach (left) assemble the fragmented bones of the fossil reptile *Hoplitosaurus*.

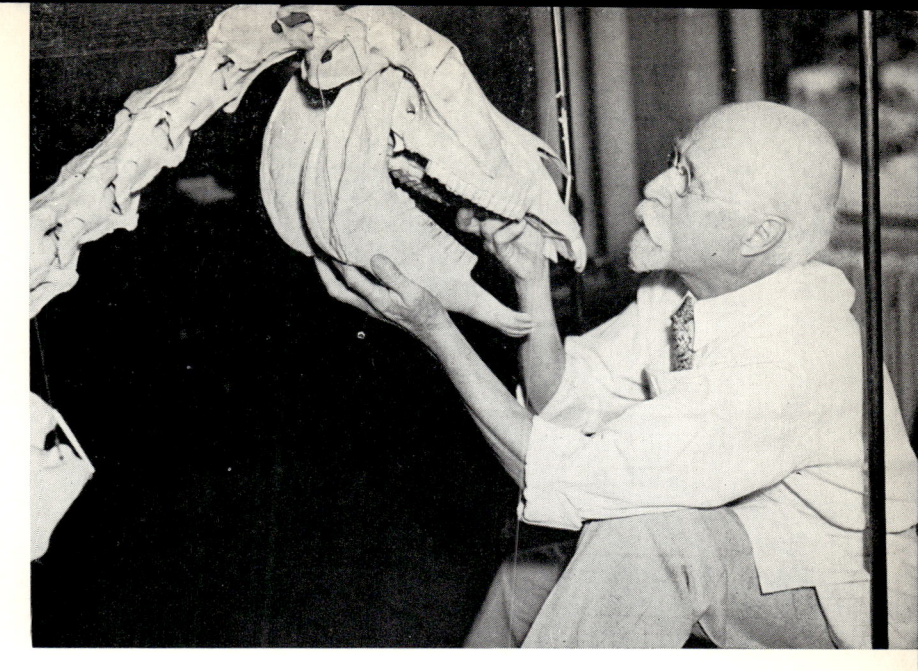

Above: S. Harmsted Chubb adjusts the jaws of a mountain zebra skeleton. This specimen and others in the Hall of the Horse Under Domestication are noted for their accuracy and beauty.

Below: Small fossils often have to be worked out of the rock matrix in which they are embedded while viewed under a microscope. Preparator Albert Thomson uses a binocular microscope while extricating the jaws of a tiny Paleocene mammal.

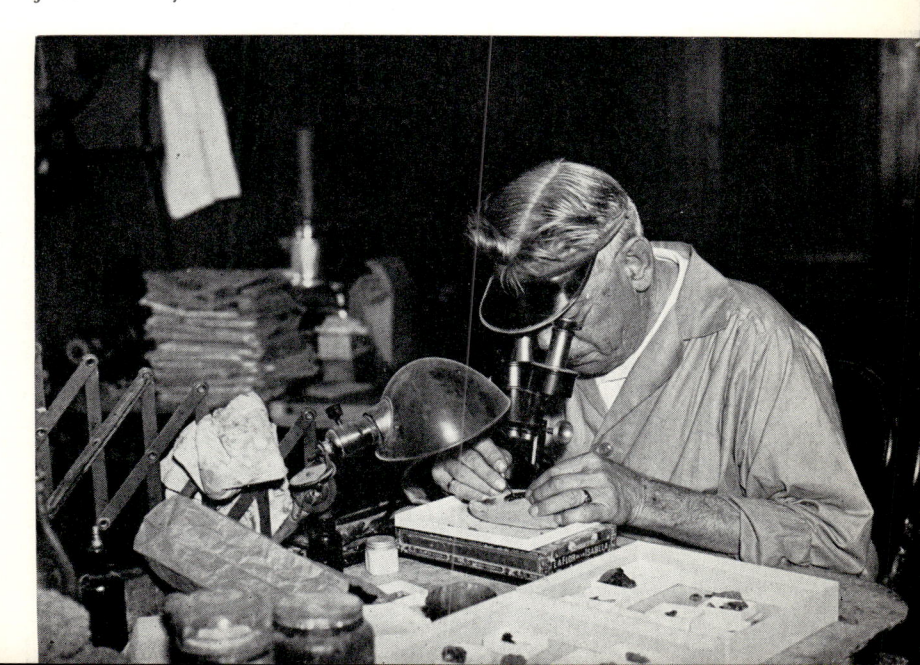

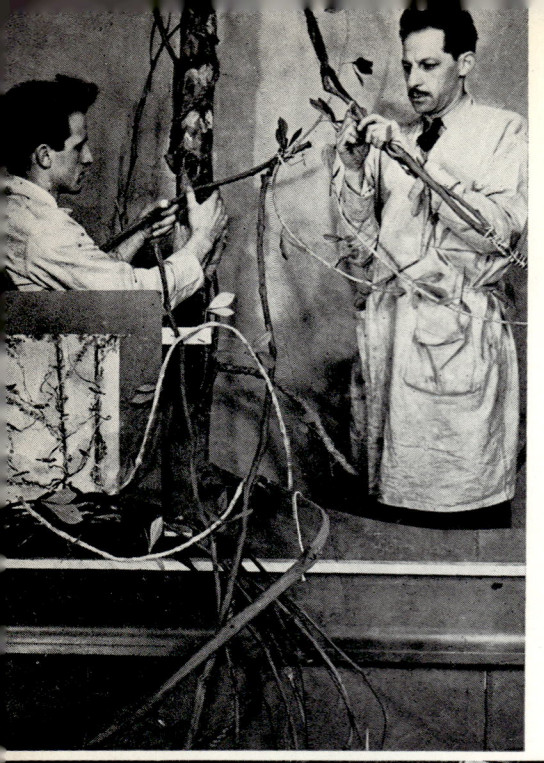

Left: **Preparators Fred Scherer (left) and Charles Tornell (right) assemble the branches of a tree for one of the bird groups in Whitney Hall.**

Below: **Dr. T. C. Schnierla, Curator of the Dept. of Animal Behavior, adjusts a maze used in testing the path-making habits of ants.**

A Cashibo warrior aims a deadly war arrow in the tropical rain-forest along the Middle Ucayala River in Peru. A display in the "Men of the Montaña," a special exhibit at The American Museum of Natural History.

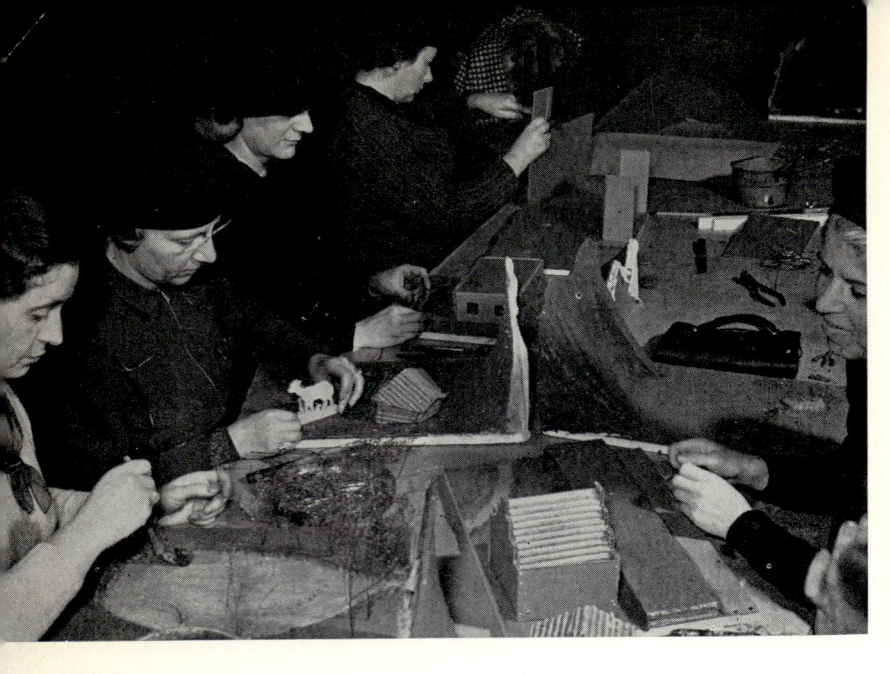

Above: Teachers go to school at The American Museum of Natural History. These teachers are learning how to make dioramas in one of the courses for teachers conducted by the Education Department.

Below: The Museum goes to School. Many educational exhibits such as the one on Indians shown here are made available for school use.

Maple Sugar in The Catskills. One of the exhibits in the new Hall of Forestry at The American Museum. This new hall will feature habitat groups of important American trees. The scene shown here is in the Catskill Mountains in New York State in the 1870's. Other groups in the Forestry Hall will include forest scenes in New Hampshire, North Carolina, and Mt. Olympia forest. The Museum's famous 16 foot wide slice of a 1341 year old Sequoia tree is on exhibit in the new hall.

"The Greatest Schoolhouse in the World." This was the apt title given to this drawing by museum artist, Christman. From all over the world, from the land, the sea, and the air the wonders of natural history have been gathered, studied, displayed and explained at The American Museum of Natural History for all to come and see and understand.

but the design is entirely characteristic of the tribe and shows no European influence. Unlike our modern art, both the Benin bronzes and the Mangbetu ivory show a strict subordination of individual expression to that of the group.

Many musical instruments such as drums, trumpets, sansas, harps, and single-stringed bowed instruments show the prominent part music and dance plays in the culture of many of the African groups. Not all of the drums were used to accompany dancers; the collection has some large signal or "talking" drums, one of which is more than six feet long and stands on four legs. A carved piece of wood at one end represents the head of an animal, while another piece of wood at the other end simulates the tail.

The drum is of the slit variety, that is, it was made out of a large piece of a log which was hollowed out on the inside with but a single long slit-like opening. Larger drums like this were beaten with mallets in order to send messages to other tribes sometimes several miles away, often to be relayed from tribe to tribe until they reached their ultimate destination. This jungle "telegraph" worked with amazing efficiency and many a white traveller has been amazed to find his arrival no surprise to the natives who were informed of his coming by signal drums long before he reached their village.

The exhibits in the hall are so arranged that the material from southern Africa is at the entrance, the material from east Africa on the east side of the hall, that from west Africa on the west side and that from the central tribes at the far end. A circuit of the hall thus give the visitor a survey of African Negro life as he would find it in making a trip around the African continent.

The early phase of most large museums is the phase of collecting. As the collections grow and as their exhibits become more extensive, their staffs turn to the problem of developing a more active interest in research both within the museum and far afield. This pattern is true of the activities of the Museum's Anthropology Department in regard to its program in general, and is particularly true of its work in the islands of the Pacific. To be sure, items from this region still find their way into the Museum but the present activities related to this area are centered in ethnological and anthropological research. Thus, today, Dr. Margaret Mead continues her investigations of primitive society in Melanesia and Indonesia and Dr. Harry Shapiro his studies of the problem of race mixture, human heridity and of the effect of environment and migration upon the human race. Dr. Mead's studies have taken her to Samoa, New Guinea and Bali, and Dr. Shapiro has done notable work on the islands of the Hawaiian group and on Pitcairn Island.

The Museum's interest in the Pacific during the first few decades of its existence was mainly confined to collecting ethnological material and in collating and studying collections which it had received through purchase or gift. As early as 1894 Mr. Rudolf Weber was sent to the east coast of Sumatra. In 1905, the Museum acquired a large collection of ethnological material from the Philippine Islands. Most of this material, representative of the life and industries of all the important tribes of the Philippines, had been on exhibit at the St. Louis Exposition of 1903. It contains, among other things, clothing and textile fabrics, household utensils, agricultural tools, fish and game traps, weapons, houses, including a tree-house, and boats.

The collection was purchased by Mr. Morris K. Jesup and presented by him to the Museum. It was further augmented by additional specimens collected by Dr. Hugh M. Smith and by a collection of swords, knives and spears presented by Mr. Charles H. Senff.

Among the interesting and outstanding items in the Philippine Hall today are a full-sized out-rigger canoe, a model of a bamboo house and a collection of the serpentine-bladed krisses or daggers. The Philippine Hall also contains material from other parts of Malaya including a display of puppets and shadow puppets from Java. Throughout the hall one notices the influence of China and India and also of Mohammedanism upon the arts and crafts of these people of Malaysia.

The South Pacific Hall which adjoins the Philippine Hall contains collections from Polynesia, Micronesia, and Melanesia. Macabre yet interesting are the tattooed Maori heads from New Zealand in the collection made by Major-General G. Robley of the British Army in the late 1860's and presented to the Museum in 1907 by Morris K. Jesup. It is the finest such collection in the world, consisting of thirty-three heads and all the implements, pigments and other accessories used in the process of tattooing and preserving. Since the practice was discontinued over sixty-five years ago such specimens are now very rare. With the Maori, tattooing, common to most Polynesian people, became a fine art, second in its quality only to the very fine work of the Marquesans.

Tattooing among the Maori was not limited to the face. Men were tattooed on the face, the upper part of the trunk, and on the thighs to the knees. Women also were tattooed, on the lips, the hands and arms, the front of the torso from the breasts to the navel and on the thighs. Tattooing was associated with the religious practices of the Maori and also with war and with rank. Women were always tattooed on the lips before marriage. The work was done by professionals who were well paid and who travelled about from village to village. So well esteemed was the craft that a slave who demonstrated ability in it was immediately freed.

The northern Maori favored the elaborate curvilinear patterns while the southern Maori kept to simple straight lines. The instruments were made of bone and wood and the pigment by burning certain kinds of wood, a variety of caterpillar, or kauri gum, in a small kiln. The soot from the fire thus made was collected on a frame of flat sticks, then scraped off and mixed with dog fat. The pattern or design was sketched on the proper part of the body with a mixture of charcoal and water and sometimes marked with a sharp point. Then the tattooer dipped his chisel-like tool into the pigment and proceeded to work it into the skin.

It was a Maori custom to preserve the heads of their chiefs, of their wives and favorite children, and those of their slain enemies who were chiefs or whose heads had been well tattooed. The untattooed warriors were left where they fell. These captured heads were paraded about during time of war and their sight was supposed to incite the warriors to greater deeds of valor. When treaties of peace were negotiated there was always an exchange of preserved tattooed heads of warriors who had fallen in the conflict. By destroying enemy heads a chief would indicate that peace was impossible.

The Maori heads were preserved by removing the brain and all other fleshy matter inside the skull. The cavities thus made were filled with flax, and the head wrapped in leaves and steamed and smoked over a fire. The smoke of the wood fire impregnated the skin with pyroligneous acid which acted as a preservative. The face was restored to its original contours by clever stuffing with flax and shredded bark. The nose cavity was stuffed with fern root and the lips were sometimes sewn together and sometimes left open so as to show the teeth.

The entrance to the South Pacific Hall is dominated by a huge monolithic head. This is a cast of one of the famous Easter Island lava rock images which was brought back to the Museum by the Templeton Crocker Expedition in 1935. It is supposed that these statues represented departed chieftains. Similar carvings were set up on stone platforms all along the coast of Easter Island, some of them thirty feet tall.

One of the gems of the South Pacific Collection is a beautiful feather war cape from Hawaii purchased by Mr. George S. Bowdoin in 1908 for $2,500.00 and presented to the Museum. It is made of feathers from Hawaiian red and yellow honeysuckers, birds so small that it sometimes took several generations to collect enough feathers to make a good-sized cape. During World War II when the Museum removed some of its treasures to a place for safe-keeping this cape was among the rare and valuable items so protected.

Several groups of life-sized figures depict natives of the South Pacific engaged in various activities. One such group shows two Tahitian women at work, one making kava, a stimulating beverage produced from the roots of a species of pepper, the other plaiting pandanus leaves into roofing material. Another group shows two Tahitian men, one engaged in grating coconut and the other in making fire by rubbing a blunt-pointed stick in a groove made in another piece of wood. The figures in these groups as well as the figure of the Tahitian priest walking over heated boulders of lava as a part of the fire-walking ceremony were all modeled after a group of Tahitian firewalkers who once visited New York City.

The islands of the South Pacific have been divided by anthropologists into five great groups based in part upon geography and in part upon the nature of the peoples inhabiting the various islands. The group nearest the Malay Peninsula and the Asiatic mainland, which includes Borneo, Sumatra, Java and Bali, has been called Indonesia. Extending eastward across the Pacific are two long chains of islands. Micronesia, the northernmost of these chains includes the Marianas, the Carolines, and the Marshall Islands. The southern chain, Melanesia, contains the large island of New Guinea, the Solomons, New Hebrides and the Fiji Islands. Scattered to the east of the extreme eastern ends of Micronesia and Melanesia is the island complex known as Polynesia. Ranging from the Hawaiian Islands to the north down to New Zealand in the south, Polynesia includes in addition to these islands, the island groups of Samoa, Tonga, the Society Islands, Tuamotu and the Marquesas. Scenes on many of these islands have been faithfully and beautifully reproduced in the Whitney Hall of Pacific Birds on the second floor.

The South Pacific Natives Hall has a small collection of Australian material. The fifth group of the South Pacific islands is Australia and comprises the continent of that name and some of the islands which fringe its coasts. The last of the Tasmanians, believed to have been the first inhabitants of Australia (the island of Tasmania is one of the States of the Australian Commonwealth) died in the 1870's. The present aborigines, a more recent and less primitive people than the Tasmanians, live chiefly in central and northern Australia. They number about 10,000 and their culture is one of the most primitive on the earth today. They possess little or no clothing, have nothing that could be really called dwellings and use very simple equipment. One of their inventions, seen in the exhibit is the boomerang or throwing stick, used in hunting. If the hunter misses his quarry with this weapon, the curved stick will continue to sweep around through the air in a graceful arc and fall to earth very close to the thrower, thus enabling him to retrieve it quickly for another throw.

The natives of New Guinea are grouped together with those of Melanesia. They are a dark-brown people, usually having frizzy hair, broad flattish noses and protruding faces. They are sometimes called Oceanic Negroids because of their resemblance to the African Negroes. By our standards of comeliness they fall far below the handsome Polynesian peoples but they are nevertheless a strong and husky race. They have developed a noteworthy decorative skill. Their decorative art is closely associated with their many rituals. They have all kinds of ceremonies, ceremonies for initiates, ceremonies for the uninitiates, ceremonies bought and ceremonies sold, all making for a considerable production in plastic and decorative expression. All these rituals call for an amazing amount of paraphernalia. Most of their artistic expression finds its outlet in woodcarving and painting. Though now and then an individual artist may seem to have broken away from convention to produce something fresh and unrestrained, a close examination of most of the specimens reveals rigid adherence to stylistic canons.

Recently acquired from New Guinea are some painted clay faces, actually built up in clay over real skulls. All have a weak-chinned look because the natives discard the lower jaw when using skulls as manikins for the clay faces.

The Melanesians are not as accustomed to making long canoe voyages over the open ocean as are the Polynesians. They stay close to their own islands. They live for the most part in simple one room houses which are nothing more than a thatch covering over a light frame. Some of their houses are raised on stilts which makes disposal of their refuse easy, if not completely sanitary. Perhaps it is not as unsanitary as one might imagine because the dogs, pigs, and chickens soon do away with any of the edible refuse. The village usually contains a large building which is elaborately decorated. This is a sort of club house for the men, and women are forced by convention to give it a wide berth. Adolescent boys usually sleep in these large houses.

The Melanesians do more gardening than the Polynesians. Their gardens are an important source of food. As seen in the Solomon Islands group in the Whitney Hall of Pacific Birds, the forests are slashed down and burned over by the men. Then the women take over with hoes and prepare the ground for crops of taro, sweet potatoes, yams, pumpkins, squashes, and bananas.

These were the natives whose garden plots were overrun by Japanese and American troops in the bloody battles of Guadalcanal and the other Solomon Islands in the early days of our recent conflict in the Pacific.

THE TWAIN SHALL MEET

Stories of cannibalism have long been associated with the natives of the South Pacific. Melanesians are fairly well noted as being fond of "long pig" or human flesh. Today, cannibalism is practically non-existent but in the old days the custom was fairly widespread and was known to take place even in Polynesia. When Dr. Harry Shapiro of the Museum was on Norfolk Island in 1923 he received the report that the natives of New Hebrides had murdered and eaten several missionaries, and in 1929 when he was visiting the Marquesas in eastern Polynesia he met an old man who told him that he was the son of a chief and was therefore entitled to a finger of any victim that might be prepared for consumption. The old man went on to tell Dr. Shapiro that this morsel was considered a choice one and that he recalled that it tasted quite good.

The Gilbert Islanders exhibit shows some of their famous armor woven of heavy coconut fiber. Also shown are some of their unusual pronged weapons, set with shark's teeth and capable of inflicting a nasty tearing wound. The Gilbert Islanders were, like the Polynesians, excellently equipped with maritime skill. Up until the last war, the Gilbert and Marshall Islands, the Marianas, the Carolines and the Palaus were under Japanese mandate and visitors and scientists were forbidden to land on them. Thus less is known about their aboriginal life than is known of the Polynesian and Melanesian groups. We all know now the reason for the strict exclusion of visitors and perhaps in the not too distant future studies on these islands will reveal much important and interesting information.

The highest development of culture in the South Pacific area is to be found in the Malayan peoples who inhabit the

East Indies. The Malays are a branch of the Mongoloid race and are short, muscular brown skinned people. The reader will recall that some of the oldest known races of fossil man were found in Java. Not only was *Pithecanthropus erectus* discovered there by Dubois and later by Von Koenigswald but also Von Koenigswald's collectors brought to him early in 1941 a part of an enormous jaw. This last discovery was named *Meganthropus paleojavanicus* or "great man from Stone Age Java." When Singapore surrendered and the Japanese took over Java, Von Koenigswald and his companions were trapped. Von Keonigswald became a prisoner of the Japanese as he was in military service. However with the true zeal of an avid scientist he managed to hide his choicest specimens, some being secreted in a milk bottle by one of his friends. The Japanese treated Von Koenigswald and his collections fairly well. They did, however, take one of the Solo skulls and send it to Tokyo as a birthday gift for the Emperor. This specimen which belonged in the Museum in Java was later recovered and returned to its rightful resting place.

The Malay civilizations of Java, Bali and of Sumatra are complicated societies, highly organized with all the religious and political complexes that go with well advanced civilization. Their art and architecture is not surpassed in the East. They have gone far in the development of music and the dance. When one examines their highly ornamented and carved buildings and temples, one senses that here is a culture which has certainly derived part of its characteristics from India. At one time Java was a powerful kingdom and some of the ruins of older temples and court buildings bear silent witness to the grandeur of this civilization.

THE TWAIN SHALL MEET

No one building is large enough to house comprehensive exhibits which reveal all the cultures of the world, past and present. Certainly, the anthropologists at the Museum realize the futility of even beginning to assume such an ambitious plan. On the other hand, the American Museum of Natural History has gone far in its anthropological researches and its collections and exhibits tell an extensive story of man on the earth. East is East and West is West, but in the studies and exhibits of the American Museum, though the West is better represented than the East, the twain do meet and demonstrate that one area is no less important than the other if we are really to understand the race of all Mankind with which we share a pretty small planet.

CHAPTER NINE

Men of Science

THE AMERICAN MUSEUM of Natural History has been popularly called a "College of Discoverers," a "Center of Information," and an "Idea Factory." Once a little boy was taken to the Museum for the first time by his school teacher. That evening his father asked him how school had been that day. The youngster's eyes lighted up and he gleefully responded, "Daddy, we had the most wonderful time! Teacher took us to the Dead Circus." On still another occasion a New York taxi-driver said to a Museum employee who had to direct him to the institution, "Gee, Lady, I didn't know this was a museum. I always thought it was an orphan asylum. Every time I drove by I saw kids going in and out."

A museum can fulfill part of its function by merely equipping a building with exhibits, but this usually results in a slow death for the institution. To stay alive and useful, it must be active in research as well as in exhibition and education. The American Museum of Natural History has been exceptionally fortunate in the calibre of the men and women who have served and who still serve on its staff.

The scientific staff is preponderantly masculine. In the early days, when collecting was the big job at hand, the rugged and hazardous work demanded much in the way of sheer physical strength and endurance, but this does not necessarily appear to be the only reason for its traditionally male staff. After all, in the middle '80's it was still largely a man's world and women played little or no part in the early collecting trips and expeditions. The situation has changed slightly in more recent years, and undoubtedly will, as time passes, change much more.

One of the earlier feminine staff members of the Museum was Mary Cynthia Dickerson. Her interests and achievements ranged from botany to amphibians and reptiles, and in her work in these fields she left her mark on the Museum. Today, although men still outnumber women on the scientific staff, we find such active and capable researchers as Dr. Margaret Mead and Miss Francesca LaMonte filling important positions in the Anthropology and Fish departments. There are other women on the Museum's research associates staff such as Dr. Libbie Hyman whose achievements in comparative anatomy are well known. Throughout the Museum, in its service departments, in its Library, in its Education Department, and in its business offices, the ladies are holding their own with the men.

It takes the cooperative efforts of the entire personnel of the Museum to make it operate successfully and it is impossible to tell of the particular contributions of each of the more than five hundred men and women who now constitute the active working force. To select a few case histories of some of the typical "men of science" who have helped the Museum to attain its high rate of achievement and service to mankind exposes the writer to the danger of committing the sin of omission. In the 82 years of its history, literally thousands of men and women have served to advance science in its laboratories and in the field. Out of these thousands, a few cases have been selected, not on the basis of relative importance, or because of force of personality, but simply because their careers may contain elements that are typical of the characteristics of museum scientists as a group.

The Museum is a big business. Its activities call for a system of administrative officers not unlike that found in a business concern. Today, the dual problem of scientific administration and business administration is handled by the Director, two Assistant Directors, and several other administrative officers. The directorship is the chief office in the administration but the vast complexity of business activities would make such a burden more than one individual could be expected to handle.

Dr. Albert Eide Parr, like many of his predecessors in the directorship, is a scientist, a marine biologist, who has achieved considerable rank in his field. Generally supervising the Museum's affairs, Dr. Parr must also concentrate his attention on the scientific and exhibition program. Mr. Wayne M. Faunce, as Assistant Director, has the bulk of his responsibility in the area of business aspects. Mr. M. F. Harty is Assistant Director for Program Administration. The Scientific and Education Departments are staffed with Curators of various rank.

The Museum now has eleven scientific departments and two divisions. In order to keep abreast of present-day trends in natural history, two divisions have been established, one of Conservation and Use of Natural Resources under Coordinator Richard H. Pough, and the other of Biogeochemistry under Consultant G. Evelyn Hutchinson.

The word "curator" means caretaker. In England, the preferred terminology has been "keeper" rather than curator. Traditionally the two words mean the same in regard to museums. They go back to the day when all that the responsible scientific officers of a museum did was to care for collections. Today, of course, the job of "caring for the collections" is but one among the many functions of curators.

The late Dr. Henry Fairfield Osborn acted both as curator of the department of Vertebrate Paleontology and as President of the corporation's board of trustees. His contributions in both of these positions were monumental and are deserving of consideration even though his story is not as typical as that of his colleagues.

Osborn was born about ten years before the Museum was founded, a son of William Henry Osborn and Virginia Reed Sturges Osborn, in Fairfield, Connecticut. When he was still a youngster his father built a small house called "Wing-and-Wing" on a steep hillside near Garrison, New York. This picturesque spot, later to be the site of the Osborn's main dwelling "Castle Rock," overlooked the Hudson River opposite West Point. Amid this relatively wild and unspoiled environment young Osborn showed an early interest in natural history. While at Princeton University he came under the influence of the famous Dr. McCosh. At first the subject of philosophy fascinated the young man, and between 1880 and 1884 he published some brief papers on philosophy and psychology.

Old World scientists were creating quite a stir in America in the latter half of the 19th century and just as Agassiz had been imported to Cambridge and there served to influence his branch of natural science in America and to train some of the men who later were to lead their field, so at Princeton, Professor Guyot, the geologist, left his mark upon American geology. Osborn was introduced to the subject of paleontology while studying under Guyot. In 1877, when only twenty years of age, Osborn and his young friends, William Berryman Scott and Francis Spiers, Jr. made a successful expedition to the Bridger Basin in Wyoming under the auspices of Princeton University. Here they found a number of fossils of little known Eocene mammals. After his return from this expedition, Osborn worked as a graduate student for two years and stayed on at Princeton for another year as an instructor.

Even at this time Osborn felt that paleontology would supply the direct evidence for evolution and that it, embryology, and comparative anatomy, would be the three main supports of the theory which was then a burning topic at all the universities.

As a part of his work during his two graduate years, young Osborn travelled to England to study under the great embryologist Balfour. Preparing for his later studies in evolution, he took advantage of his sojourn in England to attend the University of London in order to work in comparative anatomy under that inspiring teacher, T. H. Huxley.

At this stage of Osborn's career in science, it might be said that his principal interest was comparative neurology.

His studies on the origin of the *corpus callosum* of the mammalian brain and the principal fibre tracts of the brains of amphibians have been regarded as pioneer contributions in comparative neurology. This line of investigation was to be carried on by others including Professor C. J. Herrick who acknowledged the importance of Osborn's early studies in the field. Osborn even intended at the time to write a general treatise on the subject but bit by bit he increased his interest in vertebrate paleontology and by the year 1889 he had completed his memoir on Mesozoic Mammals.

It was in his early studies of fossil mammals that he traced the evolution of the mammalian molar tooth, following Cope's theory of the origin of the tritubercular molar. Paleontologists have had to work extensively with the teeth of fossil creatures. Being harder than any other part of the organism, the teeth often remain preserved after all other parts have disintegrated. Also, as animals change their feeding habits, their dentition undergoes a change. Thus the study of animal dentition is an important part of the paleontologist's quest to trace the origin of certain creatures. Osborn contributed to this field of study by devising an ingenious system of naming the major cusps or prominences of the upper and lower molars. This system was adopted by paleontologists all over the world.

In 1891, Seth Low, then president of Columbia University and Morris K. Jesup of the American Museum of Natural History succeeded in enticing Osborn to come to New York City. He was to become the Da Costa professor of zoology at Columbia and also to become the curator and founder of the Department of Vertebrate Paleontology at the Museum. At Columbia he developed a strong and progressive department of zoology with the assistance of such men as E. B. Wilson, Bashford Dean and Oliver S. Strong. While establishing the program for Columbia's zoological work he founded the Columbia University Biological Series. His book *From the Greeks to Darwin* was published in this series. There were quite a few books written at that time on the subject of evolution. Osborn's attracted wide attention. In addition, he arranged for the publication in this series of several other important works including Willey's *Amphioxus and the Origin of the Vertebrates,* Dean's *Fishes, Living and Fossil* and Calkins' *The Protozoa.*

Columbia University appointed him as Dean of the School of Pure Research, a post he held for almost ten years. When Columbia was ready to move to its present site on Morningside Heights, Osborn was given the task of planning the lecture rooms, offices and laboratories of the department of zoology in Schermerhorn Hall. During this time, he was teaching courses in neurology, comparative anatomy and paleontology. Many of his graduate students became leaders in their field. Men such as Gary N. Calkins, J. H. McGregor, Henry Crampton, and William King Gregory began their serious studies in the fascinating field of evolution under his guidance.

About 1903, Osborn began to transfer the greater part of his University lectures and laboratory work to the American Museum. His principal courses at the time were "Evolution of the Vertebrates" and "Mammals, Living and Fossil." These courses were conducted for many years at the Museum as part of the graduate program in zoology of Columbia University. In 1907 Osborn turned over this responsibility to Dr.

W. K. Gregory, his assistant and former student. He terminated his formal teaching work at Columbia in 1910 and was appointed Research Professor in Zoology. He turned back his salary for this post to the zoology department of the University and from that time on devoted the major part of his activities to the development of the Museum's Paleontology Department, to his own reseaches, and to his new duties as President of the Museum, to which office he was elected in 1908, following the death of President Jesup. Osborn served as Museum president for 25 years retiring in 1933.

When he died he left the amazing record of having written some 940 articles, papers, books, monographs and other scientific contributions. Dr. William K. Gregory lists some of his most important scientific contributions as follows:

1. The pioneer studies on the origin of the *corpus callosum* of the mammalian brain;
2. The memoir on the Mesozoic Mammals;
3. The many contributions to the study of the evolution of mammalian molar teeth;
4. The many important papers and monographs on fossil rhinoceroses, horses, titanotheres and proboscideans;
5. The great textbook on "The Age of Mammals."

In addition to these achievements Osborn also left his mark on the study of evolution by contributing certain outstanding principles or laws in regard to the evolutionary theory. These included Osborn's law of continental and local adaptive radiation, his laws of *homoplasy, tetraplasy, alloiometry, polyphyly* and *rectigradation*. The American Museum of Natural History's interest in this field is still a most active one as is witnessed by the recent publication of Dr. Gregory's "Evolution Emerging" and the important works of Dr. George H. Simpson and Dr. Ernst Mayr on evolution and systematics.

Great physical expansion of the Museum occurred during the twenty-five years of Osborn's administration. Seven immense building units, including the African Wing and the Theodore Roosevelt Memorial Building were erected in this time. During his presidency the Museum approximately doubled its size. Field parties travelled far and wide during his administration. Perhaps the period of the 1920's saw more expeditions out to the ends of the earth than ever were sent out in any other decade of the Museum's history. With all of this vast program Osborn managed to keep his balance and by careful budgeting of his time and energies and with the invaluable help of a brilliant staff he advanced the Museum both physically and scientifically to a position of high repute. He had the ability of being able to concentrate intently upon one thing at a time and thus was able to give his best to each of his many activities.

With all the busy details of his scientific and administrative life he gained much real pleasure from his interest in young people. In moments of relaxation he liked to wander about the Museum's halls observing the actions of the visitors. The writer remembers one such occasion when Professor Osborn "took over" a class of sixth grade youngsters who were studying the exhibits in Darwin Hall. For almost an hour he held the children enthralled with a discourse on Charles Darwin and on that pioneer evolutionist's travels and observations of nature.

Sixth grade classes were not accustomed to delve into such an advanced study as that of evolution. This class was visiting the Museum to study harmful and beneficial insects and had been taken to Darwin Hall to examine the large models of the house fly and the mosquito. It was a tribute both to Osborn's intense interest in young people and to his ability as a teacher that he was able completely to hold the interest of so young an audience on so advanced a topic.

He died in 1935 at the age of seventy-eight and several years after his death a bronze bust was placed in the Roosevelt Memorial Building in his honor and memory. This effigy serves to remind all who see it of the great contributions to science and education made by Dr. Osborn, but a far more important memorial to him is the Museum itself in whose interests he labored for almost half a century.

Three hundred years before Christ, Aristophanes, the philosopher, once chanced to exclaim, "But of late, birds are all the fashion." Twenty-two hundred years later this statement became one of the favorite quotations of the late Frank M. Chapman. If birds today are "all the fashion" it is partly due to the efforts of this Museum curator whose writings, both scientific and popular, whose establishment and development of one of the greatest centers of bird study in the world, and whose ability to impart to millions of people a love of ornithology, has led him to be called the "Apostle of Birds."

Born in New Jersey, near Englewood, Frank M. Chapman freely admitted that he was not a particularly distinguished student at Englewood Academy. He much preferred the woods and unspoiled fields of his father's property to the classroom. Perhaps his disinterest in school studies at the time was in part a result of an experience while attending school in Baltimore. Chapman was quite young at the time and the Civil War had ended but a decade before. He was taught history from a textbook which referred to the northern armies as "the Enemy." This was hard for the son of a Union soldier to accept and long after returning to his native New Jersey, Chapman felt a hostility toward textbooks, even though they might be free from displays of sectional loyalty.

There were a number of fine fruit trees on his father's farm and high up in the branches of a cherry tree young Chapman built a simple platform from which vantage point he indulged in two of his favorite past-times, the observation of birds, and speculative solitude. Years later, when he was sixty he went down to Barro Colorado Island in the Panama Canal Zone to begin a series of studies of bird life in this American jungle land, and this winter hideaway has become famous as Chapman's "Tropical Air Castle." This was simply a return on the part of the man to the boy in the cherry tree in Englewood. More air castles were to be set up in Florida, Cuba, Haiti, Mexico and the British West Indies and here he was to keep watch with the same fascination for the interesting behavior of birds that he had first displayed as a boy.

When Chapman finished work at the Academy in 1880 he entered upon a business career as a clerk in the city collection department of The American Exchange National Bank. Bird study was a hobby in those days to be sure, but it was only that, it simply was unheard of as a profession. Chapman was not very happy about his business career and took every opportunity he could find to seek out the company of others who like himself were interested in birds. In this manner he came into contact with some of the pioneers of American ornithology. He met and grew to know such men as Elliot Coues, William Brewster and Dr. A. K. Fisher. The American Ornithologists' Union had just been formed and about this time a prize was offered for the best paper on the migratory movements of birds in various sections of the United States. The prize for the Eastern Seaboard was won by young Chapman. He had managed to accumulate nearly all his data in the evenings after work and in the early mornings while waiting for the commuters' train to take him from Englewood to New York. The winning of this prize also meant that soon America was to lose a banker and gain a birdman.

In those days "the bird on Nellie's hat" was no illusion. Fourteenth Street was the shopping center of New York at the time and young Chapman spent some of his lunch hours there, pad in hand, recording forty different species of native American birds which had been sacrificed upon the altar of feminine vanity and had become merely millinery accessories. This was a volunteer job that Frank Chapman was doing for the newly organized American ornithologists and was to become ammunition for the first broadside in their ultimately successful war against this kind of wasteful destruction of American birds.

Chapman resigned from his job at the bank in the Fall of 1886. He had first entered the Museum with a friend of his who was an amateur ornithologist especially interested in birds' eggs and who had obtained permission to work at the Museum on Sundays, a day on which the building was closed to the public, a fact which annoyed young Chapman, since he realized that it was the only day when a majority of city people could find time to learn about Nature. Today, of course, the Museum is open every day in the year. Chapman did volunteer work at the American Museum of Natural History until 1888. Just a few days after the famous blizzard of that year, he became regularly employed as an assistant to Doctor J. A. Allen, then Curator of Birds and Mammals. While New Yorkers were still digging themselves out of the snow drifts of the great blizzard, young Chapman began his professional career in ornithology in a workroom on the fifth floor of the first, oldest and only unit of the present Museum then completed.

Chapman was a trail blazer both in the field of museum display and in the scientific study of birds of the New World. One of the first things he did in bird exhibition was to set up a display of "Birds Within 50 Miles of New York City." As an amateur he had been confused and discouraged by the Museum display of an unfamiliar array of foreign birds. He realized the need of the local amateur bird students to have accurate information on the species they were most likely to see, and thus originated the local display which even to this day serves the needs of amateurs in the city area. He also was intensely interested in developing a series of exhibits which would show birds in their natural settings. The first of the series showed marine birds of Cobb's Island in their natural environment. When President Jesup examined the group he gave his approval and Chapman embarked on the long and productive exhibition program that eventually gave the Museum its fine present series of bird habitat groups.

The bulk of Chapman's scientific investigations of bird life had to do with painstaking zonal analyses of birds of Central and South America. He and his companion, Louis Agassiz Fuertes, the noted bird artist, climbed the steep slopes of the Andes charting with great care the marked faunal territorial divisions this type of topography fostered. In the course of an hour's climb in this region one could notice differences in foliage and animal life as great as those between Canal Zone and Hudson Bay. Chapman and Fuertes brought back immensely valuable data on the effect of climate, altitude and other geographic factors on the origin and distribution of hundreds of species of birds. This type of field work was only possible in relatively few places in the world and it has set a pattern which is being followed by Museum expeditions even in the present time.

Earlier, Chapman had studied bird life in Trinidad and on other islands close to the South American coast. In his studies he obtained proof that there must have been a connecting link between these islands and the mainland. This was a kind of proof geologists had been unable to obtain previously and it enabled them to chart for the first time a hypothetical definition of Caribbean land formations in the geologic past. Another pioneer investigation of Chapman's was an inquiry into the origin and distribution of the Grackle. These studies which are still to be pursued to completion were carried far enough by him to indicate that the Grackle, a bird which today occurs in several different hues was once only purple in color. The changes in color which took place appear to be the effect of ecological factors which resulted from the marked changes that have taken place on the earth since the glacial period.

Chapman's now classic *Handbook of Birds of Eastern North America* served for years as the "bible" of local bird enthusiasts.

On October 27, 1928, Frank M. Chapman was awarded the Roosevelt Medal. Fellow recipients of this honor on that date were Charles Evans Hughes and Charles A. Lindbergh. The medal, awarded for distinguished service, was given not only for specific attainments, but even more for Chapman's "influence in the development of character." Whereas Nature had once interested thousands in America, it now interested millions and "Frank M. Chapman, by his life-long advocacy of the birds, had figured preeminently in bringing about this profound change in our habits of mind." Chapman has been called the most influential ornithologist since Audubon.

Not all of the Museum's men of science are of the "home-grown" variety such as New York Stater Osborn and Jersey-ite Chapman. Archaeologist Nels Christian Nelson was, as one might easily assume from his name, a Dane. Like some of his colleagues, he brought to the Museum a talent that was to see its highest development in the land of his adoption. Nelson has now retired but his story reads like that of an Alger hero. As a farmboy in Denmark his early concept of America was colored by the fact that some of his relatives had already immigrated to try their fortunes in the New World and by the fact that he had read and been thrilled by the novels of James Fenimore Cooper. He first became interested in Indians when he devoured Cooper's tales of the early colonial days in America and this interest was to develop into a fondness which he has retained all his days. Like many other "greenhorns" Nelson half-expected to hear warwhoops and to see real redskins when his ship docked in New York in 1892. He had been sent a steerage ticket to America by an aunt who was living in Minnesota and so determined was he to leave all of his Old World life behind him and start afresh in his new home that he flung most of his clothes out of a porthole while the ship was passing through the Narrows keeping only enough to satisfy the simple demands of decency.

America is a melting pot but in some parts of the country the cauldron seems to be set over a very slow fire. When Nelson finally got to Minnesota he found that everyday life in the little Danish community where his relatives lived was hardly different from that he had left behind him across the Atlantic. He began to work hard at the same old chores he had left behind, in order to pay back his passage money, and he realized then and there that if he were to become Americanized he would not be able to do it in another Denmark in Minnesota.

He left his uncle's farm and hired out to another farmer and started to go to school. Working and going to school at the same time is quite a task, yet Nelson completed grammar and high school in five and one-half years. It was not easy for a seventeen-year-old boy to stand up in a country schoolhouse spelling out simple words in a new language while in the company of a room full of knee-high tots who had long since learned to speak the language Nelson was struggling to master.

Since Nelson had shown such diligence and perseverance in his studies his friends tried to persuade him to become a clergyman. At first he was inclined to agree, but then, as he put it, he struck a snag. The snag was science. When his high school curriculum included physics, chemistry and botany the young man began to veer from the classics and turn to the laboratory. Working his way to California on a freight train as a stock tender, he decided to go to Stanford University. In order to get enough money to pay his tuition and board he worked as a mule-team driver and then as a hog-butcher. He wanted particularly to attend the university to set at rest some doubts he had begun to develop when he was back in high school. Some of the things he had learned in science did not seem to him to coincide with some of the standard theological explanations of the universe to which the Minnesota clergy had introduced him.

Thus he soon delved deep into philosophy at Stanford and when his professor left Stanford to take up a post at the University of California at Berkeley, Nelson transferred to that school. The more he read of philosophy the less certain he became about entering a theological career, and in 1906 he abandoned his ministerial ambitions in order to devote himself to his newly found interest in Anthropology.

It all happened by chance when he went with a friend of his to help dig up an old Indian cemetery near Ukiah, north of San Francisco. After this trip Nelson signed up for all the anthropology courses he could take. It could well have been that this interest which suddenly flowered in 1906 was an innate one which had first been stimulated when he had thrilled to the tales of Leatherstocking and Chingachgook, and which had been fanned almost to a blaze some ten years before he dug up his first artifacts, when he had been impressed by an exhibition of the history of man's tools he had seen at the Omaha Exposition in 1898. It had not occurred to him then that a man could earn his living and satisfy his desire for self-fulfillment by studying the distribution and evolution of potsherds, arrowheads and other mute evidences of the past.

Working hard in the field and the laboratory Nelson gained admission to the Museum of Anthropology which the university maintained and, after he graduated, became an Assistant Curator there. His master's thesis was a report on the archaeology of the entire San Francisco, San Pablo and Suisun Bay shore. He had a pedometer and before his survey was over it recorded that he had walked over 3,000 miles to locate and study the 400 shell mounds which his report described. Most of the time he was studying he also was earning his way by working in a local bank.

His rigorous routine got him out of bed at five in the morning. One morning, just after he had been wakened by the alarm clock, the bookcase in his room toppled over with a crash and the plaster on the walls and ceilings began to crack and crumble. It was the San Francisco 'quake of 1906, and although for a time Nelson thought that his career as bank employee and student archaeologist would be cut short almost at its very beginning, he managed to come out of the disaster without injury.

It was through Dr. Pliny Goddard, a noted west coast anthropologist and later an American Museum curator that Nelson came east to New York to become Assistant Curator of Prehistoric Archaeology at the American Museum of Natural History. This was in 1912 and Nelson, whose love was American archaeology, soon found himself excavating in Castillo Cave in Spain and visiting other important archaeological sites in Europe. Dr. Osborn's interest at the time was in Old World archaeology and he was insistent that the newly acquired assistant curator work over these European sites. Whenever the opportunity presented itself Nelson worked at American sites, in Kentucky, Missouri and, most productively, in the southwest.

When the Central Asiatic Expeditions under the leadership of Roy Chapman Andrews went into the Gobi in 1925 Nelson went along as archaeologist. This was a new continent for him and he was out only a few days with the main caravan when he made his first archaeological discovery. But Andrew's plan of march called for a rapid trip that did not allow much time for the slow, tedious and ponderous type of stratigraphic excavations archaeology necessitates. Nelson made many finds but did not really have the opportunity to do the thorough type of work he was able to do in America.

Life on the expedition to Asia was not only swiftly paced but hazardous. The Mongols were not always happy about the white intruders who were invading their hunting grounds, and more than once the truculent nomads had to be shown that the expedition members were also armed and ready to defend themselves if necessary. On one occasion when the situation appeared very serious and it looked as though the Mongols would actually make trouble, Nelson saved the day through completely overawing the natives by pretending he was a magician. His baffling display of magic consisted of removing his eye (an artificial one) and putting it back in place again.

Accompanied by his wife, he made a long upstream journey in a Chinese junk up the Yangtze River for the purpose of exploring the many caves which dotted the steep sides of the Yangtze gorges. His Chinese crew lived in constant dread of being shanghaied by the local military groups with which China was overrun, and although the Nelsons made their journey safely they learned later that after they had finally left the junk it was set upon and robbed and most of its crew were taken prisoners.

In his hunt for caves that might contain evidences of Stone Age Man, he would try to put himself in the place of a prehistoric househunter and when he saw a place that he might have chosen for a dwelling, that was the cave he then examined carefully. He would scramble up the steep bank to look inside the cavern, and, if there was any soil at the entrance of the cave, he would carefully dig for artifacts.

In theory, it is assumed that the oldest artifacts are those which are at the lowest level of a series of layers of earth which contain relics of bygone civilizations. The stratigraphic method of excavation is not very old. There was a time when artifacts of prehistoric man were simply regarded as curiosities, and as objects that had been deposited on the earth for some divine reason or other. No connection was made in pre-Columbian times between these articles and man himself. But as travelers from Europe began to encounter living races of primitive peoples in America, in Africa and in the Pacific Islands, it was as though man was seeing himself as he might have lived in earlier stages of his cultural development. In the gradual realization of the fact that European Man had himself once been uncivilized lay the germ of the science of prehistoric archaeology.

The stratigraphic method of archaeological excavation was tried out in the American Southwest by an American Museum archaeologist named George Pepper. Most people thought that this method would work in America but this hope received a setback when Pepper reported that his diggings revealed a hopeless confusion of both recent and ancient artifacts at all levels. Clark Wissler, the Museum's eminent curator of anthropology, was not dismayed by Pepper's disappointing report. He still felt that the stratigraphic method would work in America and dispatched Nelson to the Southwest to investigate the situation. After a few months work, he solved the mystery. Some recent Indians had disturbed the site that Pepper had worked and had thus ruined it for archaeological purposes before he had started to dig. Pepper, of course, had been unaware of this. Nelson chose another site, near the one Pepper had found so disappointing and soon was able to show that, in an undisturbed area, the stratigraphic method worked to complete satisfaction in the southwest. Thus Nelson has the distinction of being the first archaeologist to apply this principle of excavation successfully on the American continent.

Nelson made another notable contribution to his science by developing what is known as the "Age and Area" hypothesis. In essence this simply means that the oldest inventions have spread the farthest. Assuming that a given site is the "capital" or point of origin for a particular invention, say a style of pottery or a type of weapon, then, as one proceeds outward from that "capital" site, one may find the same invention but in more recent cultural layers than that of the site of its origin. By tracing such evolutions of a particular invention both in their vertical sequence at the mother site and in their horizontal sequences in progressively more distant sites one can more or less indicate the outward spread of a given culture by trade, conquest, or other means of communication. Nelson verified this hypothesis in his work and it was soon recognized by anthropologists that the "Age and Area" method would apply not only to particular cultures but also even to cultural development and spread in entire hemispheres. Nelson's work in the field was supplemented by equally careful studies in the laboratory. Like most retired scientists he has not retired from his pursuit of information concerning the life and customs of early man. He and his former chief, the late Dr. Clark Wissler, were in a very real sense leaders in the development of their branch of natural science in America.

The late Walter Granger, himself one of the finest field workers ever to represent the American Museum had this to say about Nels Nelson, "No man ever had a better field companion than Nels. If I were going out on a desert island where I would have to be satisfied with the company of only one man, I'd choose him." This from a fellow scientist is a high tribute indeed.

CHAPTER TEN

Operation Expedition

A MAN in a backyard in Ramsey, New Jersey, deftly catching insects in a butterfly net; a slowly plodding camel caravan moving across the shifting sands of the Gobi Desert in Mongolia; a sturdy ship crunching its way through the ice floes of the Arctic Ocean; a lone traveller sitting on a log, softly playing a tonette in a village clearing in the primitive Lacandon Forest in Mexico—these are but moments in a few of the nearly one thousand expeditions the American Museum of Natural History has sponsored in its quest for knowledge. If one were to take a map of the world and were to start pinpointing the locale of all the Museum expeditions since the first official expedition of that institution in 1887, one would find that by the time the task had been completed there would hardly be a single significant area on earth that had not been visited by the Museum's field parties.

OPERATION EXPEDITION

The aspect of Museum work that has been more successful than any other in catching the interest of young aspirants to the profession has been that of "going on an expedition." To the layman, there appears to be a special brand of glamour surrounding the very word "expedition." Year after year the Museum receives hundreds of letters from men and women, young and old, asking the same old question, "How can I get to go on a Museum expedition?" The Museum's own staff members who have gone on many expeditions are the last to glamorize this activity. So inured are they to the comings and goings of their colleagues that a situation that occurred to Dr. Barnum Brown becomes completely understandable. Brown had just completed a five-year exploration of Africa, the Near East, and the Far East. On the morning of his first day back at the Museum he boarded the elevator at the ground floor of the Museum. As he did so, another Museum curator glanced up from his newspaper and murmured, "'Morning, Brown," and started to go back to reading his paper. In a moment or so, he looked up again and said cautiously, "Been away, haven't you?"

Expeditions come all sizes. Some consist of one-man outfits and others may entail the services of a score or more scientists, artists, and preparators. Some expeditions may cover small areas, some cover a wide expanse of territory. Often an expedition may be in the field only a few months; on the other hand, the objective may necessitate the spending of years in the field. An expedition is not a haphazard and opportunistic venture nor is it only concerned with the mere collecting of things. Before one takes to the field there must be months and sometimes years of careful planning. The objectives must be set and a program of operation worked out. If all an expedition meant was to poke around into the far corners of the earth then it might be thought of as simply an adventure. When one goes into the field to survey carefully the fauna and flora of a given locality, to discover what life has been like in the past in that region, to ferret out the significance and important interrelationships between animal and plant forms and their environments, to delve into the behavior of primitive peoples, to trace the migrations of living forms from one part of the earth to another, then, and only then does adventure become real exploration.

The nineteen twenties have been called by some of the Museum's scientists the "Golden Age of Exploration." The uncertain conditions that prevailed abroad during the latter part of the '30's and World War II brought about a considerable diminution of field trips to the Pacific area and to the Far East and Africa. Now field activity is once more on the increase but until the world becomes more settled the chances are that it will be some time before the Museum can approach its record of sending 70 scientists and specialists into the field in one year.

Among the largest of the expeditions to venture forth from the Museum were the Jesup North Pacific Expedition, the Eastman-Pomeroy African Expedition, the Fossil Collecting Expedition to western North America, and the Central Asiatic Expedition. Many others of various sizes have been sent out, some with limited objectives and others with programs that involved several years of continued exploration. As an example of a very ambitious field program involving a large contingent of scientists and assistants, the Central Asiatic expeditions of the 1920's stands almost alone in the immensity of its program. At the other extreme in size, but not in importance are the small, one-man projects such as those represented by the exploits of E. T. Gilliard in the South Pacific Islands and those of Dr. Theodore Schneirla in Central America.

To give the reader a glimpse of Museum expeditionary procedure three examples will be described briefly.

It was in April, 1922 that a small fleet of five automobiles carried six American Museum scientists out into the Gobi Desert to seek evidence in support of a theory that this ancient Mongolian terrain was the homeland and dispersal center for most of the mammalian life of the world, including man. Twenty-two native helpers had left Peking with a caravan of camels several months before in order to establish depots of gasoline and oil at predetermined spots, for Roy Chapman Andrews, the leader of the expedition, was proceeding on a theory of his own, namely that the little-known Gobi Desert could be traversed by motor cars. In addition to setting up supply depots, the camel caravan would also rendezvous at appointed times with the motorcade and load up with the items collected in the field by the explorers. These rendezvous were arranged and completed by a kind of dry land navigation. Except for a few caravan routes of very ancient origin, there were absolutely no roads or anything resembling roads once one moved out onto the great elevated basin known as The Gobi. This expedition of 1922, and those which followed it annually for eight succeeding years, were almost military in organization. A small, well organized army of geologists, paleobotanists, paleontologists, archaeologists, zoologists, and photographers fought their way through sandstorms, hordes of bandits, put up with such difficulties as a 40 degree overnight drop in temperature, flat tires and motor trouble a thousand miles away from the nearest garage, and a treacherous terrain that might change in a moment from gravel desert packed hard enough to enable a car to go over 50 miles an hour to soft clay, sand, or mud that would mire a car to its hubcaps.

They operated only in the warm months of the year as the Gobi winters are extremely cold. Each year the motor and camel caravan would return to Peking in the fall. Headquarters were established in Peking and leader Andrews would often return to the States to raise additional funds for the next year's trek.

Such a life of travel was not new to Dr. Andrews. Up to 1934 he had never spent 12 months in any one country for 25 years. Inspired by the example of Frank M. Chapman, Andrews with no museum training except in taxidermy, but with a great love for the outdoors and an even greater desire to follow in Chapman's footsteps and carve out a career for himself in museum work, applied for employment at the American Museum and so insistent was he that the Director gave him a job in the Preparation Department. Here Andrews mixed clay, did odd jobs and even washed floors as he had assured Director Herman C. Bumpus he would be glad to do if this would give him the opportunity of becoming a member of the Museum staff. His first field trip was with James L. Clark to get the skeleton of a whale which was half-buried in the sands of a Long Island beach. One of his first preparation jobs was to help in the construction of the life-sized model of the giant Sulphur-bottomed Whale that still hangs on view at the Museum. Andrews was further commissioned to carry out a one-man whaling expedition to British Columbia and this began an eight year exploration of the lives and loves of the Cetaceans. His studies on the whale resulted in his receiving a Master's Degree at Columbia.

Excellent as had been his practical and academic experience in zoology, even more practical and rewarding was the experience he was to gain in the science of handling men, of getting along with them while under strenuous and hazardous conditions.

His first trip away from the Western Hemisphere was to the East Indies when the United States Navy invited the Museum to send a representative on the cruise of the *Albatross* to the Pacific and he was chosen to make the trip. He collected birds, mammals and reptiles for the Museum and also was able to stop off at Japan on the way home to investigate whales in the Japanese coastal waters.

He was to learn much more in years to come about oriental life but one of his early encounters with oriental customs took place while he was in Japan for the first time. Having put up at a Japanese-operated hotel he asked if he could have a bath. The manager assured him that this was certainly possible and that the young gentleman was most fortunate since only five other people had used the water previously. Somewhat shaken by this novel version of sharing the health, Andrews had no sooner settled down into the tub when the manager entered the room with a young girl. The host explained that the maiden would wash Andrews' back. Accustomed to making quick decisions he replied with an emphatic "Oh, no she won't!" The intruders departed but in a moment the manager reappeared with another young lady.

"*This* girl will wash your back," exclaimed the host.

Almost before the words had left the manager's lips Andrews ordered them out of the room. The poor manager tearfully entreated the young explorer to accept the second girl and sobbed out that if Andrews did not accept this one, he would have to secure a *geisha*. Andrews ultimately submitted and since that day he has managed to maintain a strictly *laissez faire* attitude toward the customs of the East.

Young Roy Andrews led the American Museum's First and Second Asiatic Expeditions in 1916 and in 1918. These early ventures indicated that a much larger expedition or series of expeditions might yield the important finds that would substantiate Henry Fairfield Osborn's prophecy that Central Asia would yield vital evidence of early life on the earth. There was as usual, the important task of raising funds and a "starter" contribution of $50,000.00 from J. P. Morgan began the flow of what eventually amounted to almost a million dollars which provided for the work for almost a decade. Contributions for the Central Asiatic Expeditions came from hundreds of friends of the Museum.

The advance base was at Kalgan, a city at the Great Wall of China, the extensive and massive barricade which has separated China from Mongolia for over 2,000 years. Kalgan was chosen because it was accessible by railroad being about 125 miles northwest of Peiping (Peking). Whenever the motorcade drove through a gate in the wall at Kalgan to head out into the wastelands of the Gobi it was like leaving for another world, and when on the way back, the explorers sighted the Great Wall again they felt they had reached civilization once more.

It was on May 3, 1922 after the expedition had traversed some 250 miles out of Kalgan that they struck "paydirt" for the first time when they found their first indications of the wealth of fossil material they were eventually to discover near a salt marsh called Iren Dabasu. Here, in Dr. Andrews' own words is the story of this significant find:

"We were hardly settled before the last two cars swung around a brown earth bank and roared into camp. The men were obviously excited when I went out to meet them. I knew that something unusual had happened because no one said a word. Granger's eyes were shining and he was puffing violently at his pipe. Silently he dug into his pockets and produced a handful of bone fragments; out of his shirt came a rhinoceros tooth, and the various folds of his upper garments yielded other fossils. Berkey and Morris were loaded in a similar manner. Granger held out his hand: 'Well, Roy, we've done it. The stuff is here,' he said. We were very happy. It was the first definite evidence that our exploration in Mongolia, so far as fossils were concerned, would not be the fruitless hunt predicted by our friends in Peking." [1]

The rest of the story is well known. In the press, in popular articles, in periodicals, in scientific papers, in the series of volumes recording the scientific achievements of the Central Asiatic Expeditions and in Roy Chapman Andrews' books, the story of the decade of exploration in the Gobi is told in full. The Museum exhibits contain the rare dinosaur eggs, the giant baluchitherium, the shovel-tusked mastodons and the hundreds of other important specimens wrested from the ancient sandstones of Asia. The project was a fine example of close teamwork on the part of the expedition personnel, the native assistants, and the scientists back home in the red granite wonderhouse on Manhattan Square.

The Museum's Department of Birds has sent its scientists

[1] Roy Chapman Andrews, *Under a Lucky Star,* The Viking Press, N.Y., 1943.

thousands of miles over land and sea to gather specimens and information. From Chapman's notable work in South and Central America, Chapin's African studies, Murphy's exceptional investigations of the oceanic birds, Zimmer's studies in South America, Mayr's South Pacific ventures and Amadon's investigations in Hawaii there has been amassed valuable additions to the Museum's collection and still more valuable data regarding bird distribution and behavior. One of the younger staff members of the Bird Department has recently completed a one-man expedition to the South Pacific. Mr. E. Thomas Gilliard, Assistant Curator in the department is following in the footsteps of his talented colleagues. Gilliard is a good example of the modern museum explorer-scientist. Already a veteran of a number of expeditions including trips to little-known areas in Venezuela and to tiny Funk Island off the coast of Newfoundland where once the Great Auk flourished, Gilliard's Museum trips were interrupted by the recent war which took him to New Guinea and Bataan with the American infantry. After the war he was sent to the Philippines to collect material for the Bataan Group and also to New Guinea in 1948 to collect for the Kokoda Track Group.

In 1950, he revisited the South Pacific to make a zoological survey of the birds, mammals, plants and butterflies inhabiting the heartland of central New Guinea. It was along the Kokoda Track area that the Australians beat back the fanatical onslaughts of the Japanese and helped to turn the tide in the Pacific combat. Many of the highland valleys of the remote and savage land were to leap into importance as emergency landing fields in World War II. It was in one of these almost inaccessible valleys that civilian pilot Lindbergh was said to have made an emergency landing in a Black Widow night fighter and many readers will remember the fantastic but true story of the miraculous escape of Margaret Hastings and the two young soldiers who crashed in a C-47 into a valley thickly inhabited by the primitive and dangerous New Guinea tribesmen.

It was into this high altitude wilderness that Gilliard went in 1950. When his expedition was planned, Robert Doyle was included as assistant, and the two men were to explore the bush after having been taken to their highland base by Captain Neptune Blood who had already done considerable exploring in New Guinea. But Gilliard had to start for the bush alone as Doyle was forced to stay behind because of a tragedy in his family. Fortunately Doyle was able to join the party after a few weeks delay and further assistance was rendered by William Lamont who later took over management of the expedition's base camp on Mount Hagen.

The expedition collected 1,600 study skins of birds representing some 140 different species, twenty-eight of which were new to science. Nine hundred skins and skulls of mammals were also brought back along with 650 herbarium specimens of flowering plants and about 500 butterflies representing 32 species.

Museum expeditions today do not limit themselves to collecting only one form of animal life. After all, the opportunity of collecting in a remote and little explored area is a great one, and although Gilliard's first interest was birds, he, like other Museum men on other expeditions, seized the opportunity to secure whatever significant specimens presented themselves.

Most of the collecting on this trip was at altitudes ranging from 5,200 feet to 14,500 feet in the Bismarck, Kubor and Hagen Mountains. In addition to his collections of animal and plant forms, Gilliard took many still and motion pictures and also noted as much as he could of the lives of the natives in the Bismarck and Kubor areas.

A small twin-motor De Haviland was used to fly the party and its equipment into the interior and from there on the rest of the trip was made with a long carrier train of almost one hundred natives who toted the 2,200 pounds of expedition equipment as well as the specimens that were constantly being added to the load. The composition of such a human supply train must be carefully planned and placed. On this planning and placement may depend the very lives of the expedition members. Of course, modern means of transportation have speeded up the business of getting near the remote areas, but one cannot explore and collect and study wild life from the cabin window of a fast plane. Now, as it was in the early days of exploration, much of the trip must be on foot.

Gilliard explained the important task of forming a workable carrier-train, "There are many pitfalls inherent in such a mobile unit. First one must realize that all of his belongings, equipment, valuables, trade goods, weapons, ammunition, medicines, notes, as well as his housing, clothing, food and scientific specimens must be lifted aloft in man-sized packs. These must be carried by wild strangers, strung out over tortuous trails like a giant centipede, sometimes a mile in length. There are many dangers to be faced by the white man who leads such a carrier line. The carriers refuse sometimes to carry their loads, which weigh an average of forty pounds, over obstacles such as brush rope bridges. One has always to face the possibility of loss of irreplaceable gear through theft and accident. To cushion such losses, all gear is divided into small lots whenever possible. Up ahead must be the leader with compass, maps, weapons, altimeter, notebook, and a multitude of light but important items ranging from a ready supply of matches, stick tobacco, glass beads, newspaper, iodine and foot plasters." [1]

Gilliard goes on to explain that next in line come the guns, such as high-power, small-bore rifles, 16-gauge shotguns and single-barrelled 410 shotguns. All of these are used in collecting, each one having its peculiar fitness depending on the nature of the specimen and the distance the shot must carry. Following the gun carriers are the men toting the cameras. This delicate equipment, like the guns, is given to specially instructed natives to guard and carry. For safety sake all the guns are carried unloaded but the cameras are all loaded and ready to go to work at a moment's notice.

Next in line come the men bearing the collected specimens and the preservative material and equipment. As the specimens increase and the load of food and ammunition decrease, the nature of the load changes. To the explorer, the collected specimens are almost as important as the food and ammunition. They represent the very reason for making the hazardous trip.

The rear of the caravan is made up of men carrying bulky items such as tarpaulins, tents, lister bags, drinking water, fuel for light, rock salt in 50 pound bags, bedding and food.

"At the tail end came two trusted natives carrying bush knives and 410-gauge shotguns. Just before them were two boys with the medical gear for we knew that injured per-

---

[1] E. Thomas Gilliard, "New Guinea's Paradise of Birds," *National Geographic Magazine,* November, 1951.

sonnel would always show up at the tail end of the line."[1] Thus Gilliard describes his carrier train. When one remembers that each day the train has to be made up, checked, marched miles over the bush, broken up when camp was made, and the whole thing to be done all over again the next day, one realizes that just the task of getting in to the remote area to be studied is an arduous one.

The expedition secured a new golden-crowned bower bird which was scientifically named *Archboldia papuensis sanfordi* in honor of Dr. Leonard C. Sanford because of the great part Sanford has played in the ornithological exploration of the Pacific, and of New Guinea in particular. Dr. Sanford, long an ardent supporter of the Museum's ornithological program, will be remembered by the reader as the person in whose honor the new Biology of Birds Hall at the Museum is named. Gilliard's expedition was able to bestow another accolade upon Sanford when they named a 12,067 foot peak in his honor. It was on the slopes of this mountain that they obtained the rare, new-to-science golden-crowned bower bird.

The basic objective of the expedition was to collect. Not much time or opportunity was afforded for making extensive behavior studies. Gilliard was aware of this when he wrote, "One must face the proposition that birds of unknown species are still to be found in a few of the remote portions of the world, for example, in the Paríja Mountains of western Venezuela, perhaps in Matto Grosso, perhaps in the Himalayas, in the heartland of New Guinea, perhaps in the Philippines, and in certain other isolated places. Although we are naturally averse to killing beautiful birds, especially in this

[1] *Ibid.*

wonderful era of conservation just now dawning, we must realize that the original surveys in the remote areas mentioned require that scientific specimens be collected and identified. If unidentifiable they must be named. This taxonomic research is coming to a close. The survey of birds, begun practically in the Dark Ages, has now virtually reached completion. The sum total of species of birds discovered throughout the world will stand at a little over 8,700, give or take a few. This includes the scant two score of species which authorities predict still remain to be discovered. It is well to remember too that in bringing this survey to a close the collector must penetrate into areas of the world which for one reason or another have been bypassed; doing this usually means that although it is already the second half of the Twentieth Century he must lapse back in mode of living to pioneer days. It is not possible for him to carry out elaborate behavior studies. As a matter of fact, such studies will be made only in the event that his basic survey shows promise of interesting material." [1]

We have already told of the trips made by the past and present members of the Anthropology department. Some of its members, like Chairman Harry Shapiro literally collect *people* to the extent that they gather important data and measurements which when analysed throw light upon the fascinating study of mankind. Others such as archaeologists Bird, Ekholm, Hay and Ford delve into the buried remains of past cultures and still others such as Bella Weitzner and Harry Tschopik, Jr. deal with the material culture of living primitives and Dr. Margaret Mead explores with the keen

[1] E. Thomas Gilliard, "New Guinea's Paradise of Birds," *National Geographic Magazine*, November, 1951.

eye and mind of the social psychologist the daily behavior of various groups of people, far and near, civilized and primitive.

An expedition with quite a different objective was that of Doctor T. C. Schneirla of the Museum's Department of Animal Behavior. This was strictly a one-man project, and unlike Gilliard, he had no long train of native carriers strung out behind him. He journeyed into the jungles of Panama and into the brush and forestlands of Chiapas and Oaxaca in Mexico mainly to observe the activities of some very small insects, namely the army ants.

Schneirla is a psychologist whose work it is to study the behavior of animals and to conduct experiments which shed light upon such functions as learning, motivation, sex behavior and other aspects of life. The Animal Behavior Department is an outgrowth of the former Department of Experimental Biology which was developed at the Museum some years ago by the late Dr. G. Kingsley Noble. Noble's studies of such aspects of animal behavior as social order, the effect of endocrines on growth and sexual behavior were further carried out and expanded by Dr. Frank Beach who now is a professor of psychology at Yale.

The present Chairman of the department is Dr. Lester Aronson and he and his colleague Schneirla are not of necessity forced to carry out all of their studies in the field as are some of their fellow museum scientists. In a well-equipped laboratory on the upper floor and roof of the African Wing, one can find quite a zoo. The nature of this live animal population changes as dictated by the type of experimentation going on. One may find a score of cats and kittens being used in a painstaking study of maternal behavior and the early learning experiences of cats, or ring doves roosting in comfortable clean cages, waiting their turn to be studied. A special dark room with such intricate equipment as mazes, an ingenious observation hood, and nesting trays, houses some of Dr. Schneirla's ant wards which he has brought back for study in the laboratory.

Since Dr. Schneirla was especially interested in the great forays of the army ants, he found it essential to go directly to the American tropics where he could observe the tiny marauders first hand. His study of this insect began in the field in 1932 while on a trip which took him to Barro Colorado Island research station where his observations laid the groundwork for a twenty-year period of research. The studies took Dr. Schneirla to remote parts of the American tropics including the old Mayan territory in Central America.

One does not always require bulky equipment to study insect behavior in the field. Schneirla usually travelled light, sometimes carrying into the bush enough food for himself for the entire trip, and sometimes depending on the generosity of natives he might meet. When working in the Lacandon Forest, an area inhabited by Indians who are direct inbred descendents of the ancient Mayans, he often used a device which shows the result of his psychological training. Travelling alone, it was most urgent that he be well received by any natives he might chance upon. When he would come to a little clearing which indicated the locale of a small village he usually would find that the inhabitants had gone into hiding upon his approach. He then would move in a most unhurried and calm manner to a fallen log or stone and seating himself in this seemingly deserted spot would take out his tonette, a small flute-like musical instrument on which he was fairly proficient. In a moment the silence would be broken by the soft melodic tones of his music. Soon he would see a few dim forms in the shadows at the edges of the clearing, then, cautiously, step by step, the children of the village would come out of their hiding places. They would soon see that there was nothing to fear from this unassuming stranger who continued to sit on the log and play pleasant flute-like music. As soon as the older natives saw that their children were unharmed they too would make their appearance, and, before long, Schneirla would be welcomed as a guest and his accommodations for the night, assured, proving once more that music is a universal language.

Schneirla's work is typical of the kind of research that utilizes both field and laboratory. It proceeds step by step through the accepted phases of the scientific method; hypothecation, controlled experimentation, observation, collection and analysis of data, and finally the verification or the abandonment of the hypothesis. It is a part of the researcher's task to determine which aspects of the problem are best attacked in the laboratory, and which problems can be more adequately approached by going into the field.

The first American Museum expedition considered official was that sent out in 1887 to the Bad Lands in Montana under the leadership of D. G. Elliot, and with Jenness Richardson as assistant, to obtain specimens for the first Bison Group. There is a record of a still earlier expedition to Matto Grasso, Brazil, led by H. H. Smith, but this does not appear to have been considered an official Museum expedition. Since 1887 the Museum has sent out thousands of scientists on over 950 different expeditions. Whereas the "twenties" saw many large expeditions, the "thirties" saw the greatest number of individual expeditions as in that decade the Museum dispatched over 200 expeditions. The year 1930 was the year during which the greatest yearly total of trips was made. Over 35 expeditions from the American Museum took to the field in that year.

During the past 62 years, Museum expeditions have visited every continent and have used every means of conveyance from "shank's mare" to fast airplanes to penetrate the polar caps, the tropical forests, the highest peaks and the depths of the ocean. The Museum explorers have faced a multitude of problems ranging from minor incidents such as a flat tire to major difficulties such as being captured by Mongolian bandits. The personnel has ranged from young and relatively inexperienced men such as was Paul Siple (now Dr. Paul Siple) when he was chosen to represent the Boy Scouts of America on Admiral Richard E. Byrd's first expedition to the South Pole, to hardy, experienced and top-notch explorers such as Robert E. Peary, Vilhjalmur Stefansson, William Beebe and Donald MacMillan. It has included a multitude of the Museum's own scientists and field men typified by such present department heads as Harold Anthony, Robert C. Murphy, George Simpson, Charles Breder, Jr., Charles Bogert, Mont Cazier, Harry Shapiro and Lester Aronson.

The modern explorer's purpose has been well described by Dr. Robert Cushman Murphy, one of the Museum's top expeditioners: "It is essential that recruits should come more and more from laboratories like our own, that those who go out into the lands and waters should have their specific problems clearly before their eyes and should be equipped with the latest resources of science for attacking them. For the haphazard age of discovery is over, and exploration is no longer an end, but a means." [1]

In a sense, the entire Museum family participates in an expedition. The librarians provide the documentary resources for pre-expedition study, the skilled artisans in the Museum's carpenter and machine shops devise clever pieces of equipment, the business office takes care of the multitude of details which provide smooth financial sailing for the journey, the administrative office clears the way by securing the necessary permits and letters of introduction which serve to open doors in foreign lands which otherwise might remain firmly closed, the shipping department sees that luggage and equipment gets to the dock on time and in good shape, the garage mechanics overhaul and check the motorized equipment, and even over the coffee after lunch, Museum scientists from other departments discuss certain expedition problems on which their previous experience may enable them to disclose valuable information. The explorer may be all alone in the field but he works with the confidence that behind him is a team of hundreds of Museum colleagues who are making it possible for his efforts to be of greatest scientific significance.

This team of scientists, specialists and craftsmen stand ready to go to work when the explorer returns home. This is when the vital task of bringing results to publication and thus to the body of accumulated scientific knowledge begins. An expedition which produces no more than specimens that stay hidden in a packing crate and observations buried in

[1] Robert Cushman Murphy, "Increasing Knowledge Through Exploration," *Natural History Magazine*, Sept.-Oct., 1930.

the unpublished notes of the explorer is of no value to mankind. The Museum has a function of diffusing knowledge as well as gathering it and through the scientific and popular press, the classroom, the seminar, the radio, television, the lecture-platform, motion pictures and last, but by no means least, the Museum exhibition, the discoveries of the laboratory and of the expedition become the property of all mankind.

CHAPTER ELEVEN

Behind the Scenes

AN ICEBERG floats with seven eighths of its bulk hidden under the surface of the water. A large and active museum must of necessity conduct its vast research and preparation program behind doors marked "Not Open to the Public," and thus like the iceberg, much of it remains hidden from public view. In basements, sub-basements, upper floors and in attics are the hundreds of specially equipped rooms, laboratories and offices in which the hundreds of museum workers labor to produce the information and to manufacture the exhibitions which make it possible to fulfill the museum's function of gathering and disseminating knowledge.

Behind the scenes at the American Museum of Natural History one finds a busy community of scientists and artists, of skilled artisans and master mechanics, of carpenters, masons, painters, electricians, glaziers, tile-setters, plumbers, accountants, printers, editors, librarians, teachers, attendants, cleaners, special policemen, bookbinders, cooks and waitresses, and a multitude of other employees all of whom contribute their part to the daily functioning of the institution. Almost every vocational activity known to man is represented in the great variety of tasks which confront the Museum staff.

In order to prevent chaos the Museum has to function under the same type of carefully planned organization which one finds in a well-run big business. At the helm of this intricate organization is the Director, who works with the assistance of a capable team of fellow administrators. These men, responsible to the Board of Trustees, carry out the policies set by the Trustees. Under their general supervision the scientific, educational, and business activities of the Museum are accomplished.

Since a museum is, or should be, an ever-growing, ever-changing institution, there is a vast amount of building and alteration that must be constantly implemented. There is a tremendous housekeeping task that calls for the services of a large corps of custodial personnel.

The activities of this phase of the Museum's existence may include such a range of jobs as doing a major overhaul on one of the Museum's power plant generators or changing a burned-out electric light bulb in a habitat group case. In a subterranean tunnel one may find Museum guards practicing marksmanship in a shooting range, and, perched atop a twenty-five foot step ladder in the Akeley Hall, one may discover an attendant "dusting off" a big African elephant with a vacuum cleaner. Working up on one of the steep roofs a crew of roofers may be found mending a leaky flashing, while in the kitchen of the busy Museum cafeteria one observes the manager discussing tomorrow's menu with the chef. These and thousands of other jobs are but routine to the business of keeping the Museum's house in order.

While the task of maintaining the buildings keeps hundreds of men and women busy, other hundreds are engaged in maintaining the Museum's scientific, exhibition, and educational program. The museum exhibit of today is a far cry from the monotonous rows of stuffed birds, shell, minerals and other specimens that lined the shelves of the exhibit cases in the museum of yesteryear. The modern museum exhibit begins in the mind of the scientist. When he has determined how his story can best be told in exhibition form, he consults with the museum architectural division regarding the best possible layout of the space assigned for the particular exhibit. Quite often the first stage is a manuscript setting forth the ideas the exhibit is destined to portray, a well-executed set of plans and drawings depicting the layout of the hall or exhibit and a series of illustrations visualizing the finished project. If men specialized in several branches of natural science are required members of the staff who have specialized in those branches are called in for consultation. If it is necessary to send out an expedition to collect materials and data, a basic bit of equipment for the trip is what amounts virtually to a "shopping list."

Men are assigned to the task of hunting, trapping or otherwise collecting the animal specimens which may be needed. Care is taken to secure good typical examples. The Museum is not a "freak" show and no extra effort is expended to obtain record sized specimens.

BEHIND THE SCENES

The method of taxidermy invented by the late Carl Akeley, and further developed by Dr. James Clark and other preparators of the Museum, calls for a procedure that begins even before the animal is obtained. Cameras, both still and motion picture, are used to "shoot" the living subjects in typical attitudes in their natural setting. After the specimen has been killed, the carcass is carefully measured. These measurements are marked in the proper place on an outline drawing. Then the photographers may take pictures of the carcass from different angles. Next, the skin or hide is carefully removed, whenever possible in one piece, always with a minimum of cutting and consequent damage.

In order to preserve the skin until it reaches the Museum tannery, the inner side is thoroughly rubbed with salt. The hide is carefully folded, hair to hair, into a package. The skin soon drys to a board-like stiffness, and, provided it does not become soaked or dampened, it will not spoil. Meanwhile, the skeleton of the creature is removed, care being taken to scrape and cut off as much as possible of the remnants of flesh which adhere to it. The loose bones are stuffed into the rib cage and the whole skeleton is then wrapped in burlap. Later on, back at the Museum, these bones can be processed in the osteological laboratory of the Museum from which they emerge, white and clean with no trace of flesh or fat left to mar their smooth surfaces.

While these early steps in the taxidermic process are being accomplished, other artists and accessory men are busy collecting plant specimens, samples of soil and rocks, and other items that will be used to complete the finished group. The site for the background having been selected, the master artist sets up his easel and begins to make his field painting which later will be duplicated on a large scale to form the painted background for the finished group.

Each of the Museum's habitat groups shows not only an actual locality, but shows it at a particular season of the year and at a particular time of day on a certain kind of day. Thus, for example, the Timber Wolf Group in the North American Mammal Hall shows a scene in winter on the frozen-over shore of Gunflint Lake in northern Minnesota. The time is toward the middle of the night and even the stars that have been carefully painted into the night sky are where they should be at that time of night at that time of year. These details are carefully checked and thus make the groups truly accurate representations of nature.

An expedition which has been sent out to collect materials for a habitat group thus will bring back the salted skins of the animals to be shown in the group, a field painting of the background scene, painted by one of the Museum's master artists, the skeletons of the animals and measurement data for each creature, dried and preserved plant specimens, samples of soil and rock, plaster casts and molds of leaves and other parts of plants, carefully painted color notes for the plant specimens to be eventually shown in the group, and many photographs in black and white and in color which will enable the preparators to give the artificial representations of the foliage in the finished group their proper shape and color.

After the expedition returns, the field men now become laboratory men. They set to work to create as life-like a representation of a natural scene as it is humanly possible to do. One of the first things that is done is to make a small-scale version of the group to be constructed. This three-dimensional model serves many purposes. It helps to determine the position and attitudes of the animals in the group, it helps the artists to obtain the best possible composition of the items in the foreground of the scene and it helps to determine some of the lighting problems of the setting. Once more, any of the Museum's staff who have had experience in the region which the group depicts are apt to be called to the preparation studio to give their advice while the group is still in the small-scale, temporary and therefore changeable model state. Finally, the model is approved and work on the large group begins in earnest. The animal skins are sent off to the Museum tannery where they are properly processed.

If the skin is one of a mammal with a sparse coat of hair, such as a rhinoceros, elephant or hippopotamus, it is given a bark tan. This process stains the skin a dark brown, but since there is so little hair, the tanned hide can be easily colored back to the proper shade of gray by means of a paint sprayer. If the skin is the pelt of a furred mammal like a lion or a leopard, the bark tan method cannot be used as it would permanently and irrevocably stain and discolor the hair. These skins or pelts are given a salt and alum pickle.

Meanwhile, the taxidermist, or rather the sculptodermist, has begun work on the life-sized clay figure which will be used as a positive form from which a mold will be made. This is the "Akeley" method. At one time, animals were mounted by stretching their skins over a solid form made of wood, iron rods, excelsior or straw and clay. These forms were prepared by men who were not necessarily sculptors and the resultant mounted animal often did not appear at all life-like. More than that, the solid, but relatively insecure straw or excelsior stuffing did not hold its shape well and the exhibit-life of such a museum specimen was short. Akeley's method calls for the talents of a sculptor in the creation of the clay life-sized model. This model must check with the girth measurements which were made of the animal's carcass before it was skinned.

After the clay form is modeled, a sectional plaster of paris mold is made from it. In each of its sections a permanent manikin, often less than a quarter of an inch thick, is built up of layers of cloth, papier mâché, glue and wire mesh. The sections are then assembled and the tanned or pickled skin is draped over the finished manikin. The entire task may take as long as six weeks. Often it may take as long as a week for the task of draping and securing the skin alone. Any stitching is so well hidden by hair that often it is impossible for the layman to find it, even though he examines the finished specimen closely.

While the taxidermists are preparing the animals, other artists, called accessory men, are busy creating the foliage in the background for the group in cleverly and beautifully fabricated artificial leaves and flowers of wax, crepe paper, wire, cellulose-acetate and oil paints. As a rule, grasses, mosses, the branches and the trunks of trees are natural. They have been preserved in formalin and glycerine, or, in the case of the tree trunks, hollowed out on one side and impregnated with wax to prevent spoilage and checking or cracking. Molds are made in the field from fresh fruits, flowers, and leaves, enabling the accessory men to cast perfect replicas.

Another group of craftsmen may be preparing the ground mass of the group. Snow is often made of borax or paraffin.

BEHIND THE SCENES

Careful field study sketches and color notes as well as small samples of the actual rock enable the Museum artists to produce seemingly perfect imitations of rocks out of wire mesh, burlap, papier mâché, plaster, wax and paint. It is often quite a shock to see a preparator lift with one hand "rock" that looks as though it might easily weigh half a ton!

In the actual alcove where the finished group is eventually assembled, the master artist works on the large reproduction of his field painting. By ingeniously squaring off the curved background in charcoal lines and treating the field painting in similar fashion, the master artist and his assistant can carefully and accurately transpose the field painting to the large background.

When the background painting is almost finished and the animals and foreground materials are ready, the task of assembling the group begins. An important part of this final work is to "tie in" the foreground with the painted background. This is accomplished in several ways, one of which is to show part of a rock or a fallen log in the foreground and then complete the rest of the log or rock on the painted background.

Over each group is arranged a specially equipped lighting system with lights of the proper wattage and shading to give the finished group its realistic appearance.

The last thing to do is to place the glass front into its frame. The group is entirely sealed off except for a thin metal breather tube filled with a filter of fine steel wool. It is necessary to have these in each group to allow a "point of least resistance," which takes care of expansion and contraction within the group due to temperature changes.

The painstaking work necessary to complete these large

habitat groups means that the cost of a single one may often be as much as twenty-five thousand dollars. They are among the more spectacular of the Museum's exhibits, beautifully accurate representations of nature which continue to delight the millions of visitors to the Museum.

Every department has its problems behind the scenes which the average visitor seldom suspects. In order to prepare the fossil specimens that fill much of the upper floor of the exhibition area, the Department of Paleontology must maintain a large laboratory and workroom. Here the large burlap and plaster-wrapped field specimens are uncovered and then during months of diligent patience and hard work, the fossil bones are freed from the rock matrices which have entombed and preserved them for millions of years. Here one may find a technician using a fine dentists' drill and a magnifying glass to work out of a hard rock matrix a tiny bit of the jaw bone of one of the early ancestors of our present mammals while nearby another preparator and his assistant use a heavy block and tackle to raise into position the six foot long femur of some extinct reptilian giant. It is an extensive and expensive job to prepare the skeleton of a creature like the sixty-eight foot *Brontosaurus* that is on exhibit in the Jurassic Hall. As the skeleton is being assembled, the laboratory workers must become blacksmiths and iron workers in order to fabricate the iron and steel framework that supports the heavy fossilized bones.

The unparalleled series of mounted skeletons of horses in the Horse Hall on the fourth floor of the Museum represent years of preparation work at its best on the part of the late S. Harmstead Chubb. After a thorough study of the anatomy of each animal Mr. Chubb mounted each of these horse skeletons with such care that the visitor who beholds them is struck with their life-likeness. These are not merely bones hooked and fastened together, but skeletons put into place, bone by bone, each in the position it would occupy when the animal was in the posture selected. It often took Chubb an entire year to produce one of these "x-ray" accurate skeletal mounts.

In order to provide realistic and often enlarged reproductions of many kinds of animals and plants or of some of the significant parts of organisms, the services of such artists in glass and wax as the recently retired Herman Mueller and Vincent Narahara or of present scientists and craftsmen such as Dr. George Childs are required.

Such men as Ernest Neilson and Fred Scherer work in the Anthropology Department's studios restoring priceless artifacts from bygone cultures or building miniature groups which depict such ethnological scenes as a Hopi Indian pueblo or an Ojibway Indian Village.

In the Insect and Spider Department the task of mounting the hundreds of thousands of insect and spider specimens for the study collections occupies the time of trained assistants and the job of caring for the live insect zoo poses a feeding problem that would tax the talents of a trained dietician. Especially in the exhibits of the departments that deal with the smaller members of the animal kingdom is there the added assignment of skillfully and accurately making enlarged models of these tiny denizens. In all this work the scientist and the artist-preparator labor side by side, the one assuring the scientific accuracy of the creations produced by his talented colleague.

The problem of caring for the vast study collections of

all forms of animal life past and present, and of hundreds of thousands of artifacts representing material cultures of peoples the world around, is one that calls for a tremendous amount of accessible storage space and voluminous catalogues as well as frequent cleaning and disinfecting. For example, the upper floors and attic of the Anthropology Department's section are honeycombed with vault-like storage rooms. Stepping into one of these rooms is like stepping from New York into Africa or India, or perhaps like going backward in time to the great pre-Columbian cultures of the Mayan and the Aztecs in America. Each pot, warbonnet, weapon, dress, musical instrument or tool bears its important catalogue number and each item must be stored safely yet must be readily available for the use of museum or visiting scientists who wish to examine and study it.

It has been estimated that the Museum has in its study collection about ten times the material it has on public exhibition. The twenty-three acres of Museum floor space may seem to be more than ample to the layman, but the Museum worker knows full well that the building, vast as it is, is even now far too small to house its treasures properly. A walk through the corridors of the floors not open to the public reveals the overflow of cases, jars, and boxes, all containing valuable scientific material, which have had to be placed along corridors, and in some instances out in the open courtyards under tarpaulins simply because there was no more room left in the regular storage areas. Laboratory space is also becoming crowded and it is evident that the largest of American museums, and one of the largest museums in the world, because of its intensively active research into the natural history of the world is and has been for some years, too small.

But mere size is no indication of the greatness of a museum. Unlike the old days when collecting was a race to see who could get the most regardless of its present or future significance or regardless of its being used, the Museum today cannot afford to collect anything but what will help its men of science to write, and dramatize through exhibits, the important story of man in nature. The public sees the completed results of years of work in a printed publication or in an interesting and exciting exhibit but it seldom realizes the many and varied operations behind the scenes which have had to function to make such results possible.

A few glimpses into some of the research programs under way at the American Museum of Natural History in the last few years will indicate that this "behind the scenes" activity of the institution is keeping pace with the expanding world of science.

Oil is vital for the national well-being of our country and of all the major countries in the world. The Department of Micropaleontology at the Museum has become one of the world centers of information regarding the little-known but tremendously important group of organisms known as fossil foraminifera. Under the direction of Chairman Dr. Brooks Ellis and his staff including Associate Curator Angelina Messina and Assistant Curator Eleanor Salmon, much valuable data has been gathered on these creatures of prehistoric seas. By studying the structure of these minute forms the scientists can form a picture of sub-surface architecture and this helps in the location of geologic "traps" which may be responsible for the accumulation of oil deposits. In this manner, museum scientists influence man's very vital quest for a very important natural resource.

The Museum's work on cancer research both at its Lerner Marine Laboratory in the Bahamas and at Manhattan Square is now assisted by a grant from the American Cancer Society. This fund is being used to enable the Museum to continue the studies by Dr. Myron F. Gordon on normal and abnormal growth in fishes.

In 1948-1949 a team of research workers from the Sloan-Kettering Institute for Cancer Research spent almost a year at the Lerner Marine Laboratory working on the problem of growth-retarding substances. In the same year, the Museum's Curator Emeritus of Fishes and Comparative Anatomy, Dr. William K. Gregory, received the top award presented annually by the American Board of Orthodontists for his important contributions as an evolutionist and paleontologist on fossil and modern teeth.

During the years which followed the end of World War II in 1945 the Army struggled with the almost impossible task of identifying its war dead. In many cases the usual means of identification were lost or destroyed. Responding to the need, the Museum's Curator of Anthropology, Dr. Harry Shapiro worked out a system whereby identification could be made by a study of bone structure and measurements.

Dr. Howard Curran of the Department of Insects and Spiders recently carried on a research project in cooperation with Sperry Gyroscope Company which involved studies of the flight of the drone fly made with super-slow motion pictures. Discoveries were made in this study which will be of value in the science of aeronautics and particularly to the manufacturers of instruments designed to stabilize the airplane. Drs. Cazier and Gertsch continue their studies of insect life in the Southwest.

In another part of the Museum, Dr. Margaret Mead of the Anthropology Department is at work on a study of Soviet cultures by anthropological methods. It is a far cry from Samoa, the scene of Dr. Mead's earlier culture studies, to Russia, but it is hoped that the completed studies may provide the basis for a better understanding of the Soviet peoples.

The Chairman of the Anthropology Department, Dr. Harry L. Shapiro has recently been working on a biological study of the people of the island of Puerto Rico.

Dr. Schneirla of the Animal Behavior Department is concerned with a program of investigations involving cats, birds and ants. The Chairman of that Department, Dr. Lester Aronson and a team of investigators have been delving into the psychological and physiological aspects which underlie the reproductive behavior of fishes while in the same division of the Museum, Dr. Bernard F. Reiss has been studying the influence of experience in the maternal behavior of the rat.

Dr. Harold E. Anthony and his colleagues Dr. G. H. H. Tate, George Goodwin and T. Donald Carter continue their broad program of the classification of mammals from all parts of the world using data obtained from extensive collections of mammals from Australia, New Guinea, Mexico, Africa and other areas where its field men have explored.

The problem of thermoregulation in reptiles has recently been attacked by the Museum's chief herpetologist, Charles M. Bogert. These studies are of extreme significance to members of the medical profession who have been conducting research on the maintenance of a constant body temperature in human beings.

Hundreds of thousands of persons have been taken on special tours of the Museum by the members of the Department of Public Instruction staff. It is no small wonder that the tour which has most impressed visitors has been that which, of necessity, must be limited to a relative few, that is, a trip "behind the scenes." Here one has the opportunity to see at work a team of talented and learned men and women whose remarkable achievements are in a sense more awesome than the most spectacular of the Museum's public exhibitions. An incalculable number of lives have been influenced for good by the efforts that have gone on behind the scenes at the American Museum. That fact, the acquisition of knowledge which has resulted from the laboratory researches of the Museum staff, the contribution to human welfare and understanding the publications and exhibits and programs of the Museum have made, and the simple and unmitigated pleasure the beautiful displays have given to millions who have come to see them, are the real fruition of the plans of the men who founded the institution in 1869.

A museum is not just a building full of things. It is an idea about ideas nourished and developed and brought to reality by the museum men and women whose daily labors the public seldom sees. A museum without the staff of inspired workers behind the scenes would soon become obsolete and functionless. The American Museum of Natural History has kept alive and has kept on growing, not merely in size but in useful function because of the vision and public spirit of those who provide the financial means for its existence, and have labored in its vastly varied occupations.

CHAPTER TWELVE

Man and Nature

IT WAS a blustery, cold Spring day and the rain was falling with a steady monotony that threatened to continue for hours. Charles Russell, special assistant to the director, huddled under the protection of the heavy archway in front of the 77th Street entrance looking for New York's "needle in a haystack," a taxi in a rainstorm. A cab turned in from the street and began to descend the gradual slope of the driveway and finally pulled up before the entrance. After Dr. Russell had given directions to the Negro driver, he settled back into the seat for a ride across town.

When the cab halted for the first red light, the driver turned and spoke to his fare. "You work in that place, Doc?"

When Russell said that he did, the driver delivered the following monologue:

"For a long time I've wanted to tell someone who works

in that museum what it has meant to me and to my daughter. My little girl, she's only ten, came home one day and said that she had just seen the most wonderful thing in the world. She wanted me to leave the house right away and go to see it again with her. Her school teacher had taken her class to that museum and a museum teacher took the kids to a room where there were all kinds of beautiful things made of ivory and wood and iron. The museum teacher said that these things all came from Africa, that they had been made by people who were artists and craftsmen.

"My little girl never knew much about Africa except that it was a place where Negroes came from. Now she saw what beautiful things had been made by her ancestors in their home land.

"Then they went to a big hall full of windows with animals and scenery all from Africa.

"She could hardly wait till she got home to tell me about it. She told me she was happy and proud to see the beautiful country her people had come from and to see the wonderful things they made.

"I work long hours, Doc, but on my next Saturday off she and I went to your museum and it was like she said. I never gave much thought to Africa or to museums for that matter, but when I saw the look of wonder and happiness on my kid's face as she took me by the hand and began to tell me about our own people, I was proud and happy that we had a place like the museum in New York that could mean so much to a kid from Harlem. That's why I asked if you worked there, Doc, I just had to tell somebody in that place how much it meant to me and to my little girl."

The cab slowed down and pulled up to the curb. Doctor

Russell paid his fare and got out, feeling rather humble and a little overcome with a realization of what a privilege it was to be a part of something which could so deeply affect the lives of others.

Over 2,500,000 persons have visited the Museum in the past year. Since 1869, when its career began, probably over 100,000,000 men, women and children have crossed its threshold. More millions have been touched in some way or other by the Museum's publications, its lecture services, its radio and television programs and its vast education department extension services. To the extent that the contact has been meaningful to the individuals involved, the Museum has been successful.

At a recent meeting of the International Committee on Museums of UNESCO in London, Dr. Albert E. Parr, the present director of the American Museum of Natural History read a paper entitled "Museums of Nature and Man." He must have been thinking of the people like the cab driver when he said, "When we think of other people without knowledge of their environment, it is like thinking of them in a vacuum which we are forced to furnish in our minds with the type of environment we ourselves are familiar with from our own surroundings, if we furnish it at all. We become impatient with their reactions, because they differ from ours and we have no knowledge of the reasons why they are not the same. We may become critical or even scornful of their way of life, even though it may actually represent an accomplishment we ourselves might have difficulties in duplicating if we had to live our lives in their natural circumstances."

There has been quietly taking place a tremendous broadening of purposes in museum research which has already resulted in museums becoming highly important centers of ecological research. The ecological approach has been utilized in the newest hall in the American Museum, the Felix M. Warburg Memorial Hall, made possible through the generosity of the Warburg family to honor the memory of the late Felix M. Warburg, long a Museum Trustee and a generous supporter of its scientific and educational activities. It embodies in its exhibits and labels an ecological treatment of a given area, the region surrounding the little hamlet of Pines Plains in Dutchess County, New York.

Dr. Henry Svenson, the Museum's Curator of Forestry and General Botany, and Director Parr developed and executed the plan for this hall with the close cooperation of members of the staff in the Museum's other scientific Departments. It is the result of a joint effort of many men of science. It is the first full-sized hall of its kind in the American Museum.

Ecology may be a strange word to most people. Louis Bromfield, in an article on the new hall in *Natural History Magazine,* defines it thus: "To most people the term 'ecology' is a sixty-four-dollar word that conveys little meaning. Actually its significance is comparatively simple. It means the relationship of plants, animals, and people to their environment and the alterations that occur in the environment from time to time through changes of season, rainfall, climate, and the influence of man himself. It is a science, and basically even a philosophy, which in the past has received too little attention in our whole system of education. It is basic because only out of the proper understanding of man's relationship to his surroundings can he eventually achieve health, prosperity, vigor, the means of relaxation, and indeed not only peace of mind for himself but peace with his neighbors. It is not too broad a statement to say that if the whole of the world were established upon a base of proper ecological understanding, the occasions for wars would be greatly diminished or even eliminated."

The Warburg Memorial Hall is the first of a series of projected halls that Dr. Parr has called Man and Nature halls. The story of ecology calls for its support from all the fields of natural science which the Museum has so carefully developed in its long and fruitful scientific career. One does not need an extensive area to demonstrate its meaning. Therefore, the area of 40 square miles around Pine Plains can be used to demonstrate ecological principles of great significance just as well as an area as large as New York State, North America or the world itself.

Pine Plains is only 90 miles from New York City and can be reached by modern highways in less than three hours. It was chosen as the locale to be treated in the new hall both because of its nearness to New York and because the surrounding countryside presents a rich variety of conditions. Here are swamps and marshes, both active and abandoned farm land, ledges of decaying shale and granitic rock, mounds of glacial till, potholes, forests, undergrowth and wildflower vegetation. The entire area, although populated by humans since the day of the Indian, still maintains a goodly array of wild animal life of many kinds.

After introducing the visitor to the area as a whole with its outstanding physiographic features of Mud Lake and Stissing Mountain, the exhibits break down, as it were, the area into its components and demonstrate the many important interrelationships that exist in the area among its rocks, soils, climate, plants and animals including man himself.

The visitors' attention is called to the exhibit by a large habitat group which shows Stissing Mountain in Autumn, its graceful slopes aflame with a riot of color reflected in the waters of Mud Lake which nestles at its base. The foreground of the group reveals the presence of not just one kind of animal life but of the various species that one might find in a woodland setting. Some ducks are taking off in flight and the cause of their disturbance, a red fox, is detected stealthily moving past a large ant hill. Two blue jays have spotted the fox and are screeching down at him from their safe vantage point on the branch of a nearby tree. Closer examination of the group reveals still more animal life such as prarying mantes, monarch butterflies and a woolly bear caterpillar.

The mountain, the forests on its slopes and the lake at its foot, the small streams which feed the lake and the lush marsh at the other end, the open meadows, woodlots and tilled fields of the farmlands of the region all are the setting for the fascinating and important interrelationships of living things with other living things and with the inanimate ingredients of the Pine Plains area. The story of these interrelationships, the story of the ecology of Pine Plains, unfolds before the visitor's gaze as he moves from exhibit to exhibit.

A large diagram on a wall near the entrance calls attention to the major points of interlocking principles which have operated to make Pine Plains what it is today. These very principles can be applied to any area in the world and once clearly understood, they enable one to read the story of other regions and of other peoples. Beginning with an introduction to the geology of Pine Plains and its surrounding area, the story continues with exhibits that show the importance of water to the area, the effect of the great ice ages upon its topography, the nature and origin of its soils, and the interrelationships between certain types of soils and plants. By means of carefully painted backgrounds the visitor always is reminded that he is still considering the Pine Plains area.

Amazingly realistic cross-section groups depict life on and under the soil in winter and summer, on the farmer's lawn and near the lake shore. Vegetable life in the soil continues the series and leads the visitor on to displays and dioramas that vividly indicate the importance of understanding the soil and its management, by the rotation of crops and other scientific agricultural practices. Two great agricultural activities now are pursued in the area, dairy farming and apple growing.

Another dynamic exhibit shows the importance of two insects to the business of growing apples, the bee which pollinates the apple blossoms and the codling moth, whose larva is the well-known "worm" in the apple. Another display reveals the chronological development of natural and synthetic fertilizers by farmers in the area, from the Indians to the twentieth century agriculturists.

The importance of the natural cycles of decay and nutrition are made clear by three dimensional groups and simple but effective diagrams. A large habitat group showing a cross section of the area from a farmer's field to the lake presents not only a picture of the life, both animal and vegetable above the ground, but that of the subsurface and subterranean organisms as well. Additional exhibits develop the theme by indicating the effects of bad grazing habits and of forest fires.

Culminating the series is a panorama in which five scenes merge into a seeming single picture, yet each of the five large panels depicts the forty-square-mile area at one of the five stages of its development. First one sees the Primeval Forest inhabited by wild creatures and the primitive Indians; next, a scene dated 1790 shows the first white settlers coming into the area, their clearings dwarfed by the patches of primeval forest still uncut. Then a panel marked "The High Tide" depicts the area in 1840 when the land had been cleared of forests and put to agriculture. This was an era of agricultural prosperity for the Pine Plains area as the freshly cleared virgin soil, still unspoiled and undepleted, yielded abundant returns of produce. The inevitable next scene, "The Ebb," is set in 1870 and vividly portrays the beginning of the decline in fertility of the soil which was bound to result from an unwise agricultural practice that took everything and returned little. The fifth and last state of the area shows a scene in 1950 when, with the knowledge of an agricultural procedure that can enable men to restore land damaged by bad management, modern farmers are rebuilding the soil to its old level of production and fertility.

Thus, the important message of the hall is this: that, armed with a proper understanding of ecology, man can restore to a farmed-out land a productivity which not only can equal but often will even surpass that which it possessed when it was first put to the plow. In this new hall, we have the new concept in museum exhibition developed to a remarkably effective degree. This exhibit started with an idea, that of telling the story of man and nature in a rural area in New York. This idea was then implemented with all the scientific knowledge and artistic ability which the Museum's talented corps of scientists and display experts were able to muster for the effort. It represents another fine example of the educative force which a modern museum can be.

Not very long ago, the Museum received a letter from a schoolgirl which contained the following request, "Please send me the curator of reptiles. I need it for my homework." The young lady may have been sadly misinformed curatorially, but she, like thousands of her fellow students, was wise enough to call upon the Museum when she needed help in understanding the natural world in which she lives.

The Museum receives thousands of letters each year from people in all walks of life who have learned that it is a center of information on natural history. Some of these letters may contain easily answered questions concerning the treatment of insect garden pests; they may consist of hard-to-answer problems which come from researchers in other related fields of endeavor. No matter what their degree of difficulty, they are all indicative of the need and desire on man's part to understand his environment more adequately.

In addition to its scientific staff, its vast and informative exhibitions, the Museum maintains three other important agencies to help in the promulgation of knowledge. First, there is the Library with its almost two-hundred-thousand volumes on natural history. The librarian and a corps of trained assistants provide a service that is invaluable to the Museum's staff and to the many outsiders who daily use the library's facilities.

A good library is absolutely essential to the operation of

any active scientific and educational institution and the seemingly everlasting patience of the Museum librarians is always evident even in the trying days when some local newspaper is conducting a picture puzzle contest. The librarians have solved the rush which always results in a simple way. Each day of a particularly popular contest, they look up the strange creatures offered for identification, identify them or have them identified, and then type a list of the answers and send it out to the information desk. This may not be entirely "cricket" to the puzzle people but it does free the librarians for their far more important task of servicing the serious users of the library.

Another great medium of spreading knowledge which the Museum uses is the printed word and picture. In addition to the many scientific publications of the Museum it maintains a program of popular publication. The Museum's own *Natural History Magazine* is known the world over for its high standards of literary achievement and pictorial presentation. It has been in steady publication for more than fifty years. Much of the story of the American Museum of Natural History has been told in word and picture in its pages. The author has leaned heavily upon the articles in this excellent journal and many of the quotes in this book have come from the Museum's own famous magazine. *Junior Natural History Magazine* brings the treasures of natural history to the younger readers and the hundreds of leaflets, booklets and handbooks dealing with natural history subjects and with Museum exhibits provide readers with up-to-date information on many branches of natural history.

The Museum has not neglected the media of the public press, radio and television. Literally thousands of programs have been broadcast by Museum scientists over all of the leading networks. Many of these programs have originated in studios or at the Museum but on frequent occasions the mountain goes to Mohammed and the world hears a Museum broadcast from a plane cruising above the clouds while Museum astronomers are photographing and observing an eclipse, or from the heart of the Amazon jungle where a hardy band of Museum explorers have penetrated to study the natives in their own primitive surroundings.

Albert Smith Bickmore began this book and rightfully he comes into the picture at its conclusion. The educational program of the Museum which today is activated by the largest education department of any museum of any type in the world was instituted by Professor Bickmore. A formal program of public education came into being in 1880 when the Trustees authorized Professor Bickmore, then Superintendent of the Museum, to prepare for the public school teachers a special course of lectures on natural history to be given at the Museum and to be illustrated with the Museum's collections. The first lecture was given in 1881 and was attended by twenty-eight persons.

In 1884, the Department of Education was established and Bickmore became its first curator. This department has steadily grown and widened its sphere of influence. Since Bickmore's day it has been expanded from a one-man lecture service into a veritable beehive of educational activity employing many specialists in its several divisions engaged in both direct and indirect teaching programs and services.

The fine relationship the Museum has with the schools, a relationship that was planned by the Museum's founders, was developed by constant cooperation between the Museum and the New York City School System and is in no small part the result of the efforts of Dr. Grace F. Ramsey, the Museum's Curator of School Relations.

A museum is certainly in itself the finest type of visual aid and the Museum educational program pioneered in the field of visual aids such as lantern slides, motion pictures, and specially designed three-dimensional materials. Under the auspices of the Education Department, the pioneer outdoor nature program in the East was developed at Bear Mountain, New York in 1926, in conjunction with the Commissioners of the Palisades Interstate Park. This 57-acre tract on the site of historic Fort Clinton in the Highlands of the Hudson, with its nature trails and its seven museum buildings, its animal enclosures and beaver pond, was one of the earliest of the park museums and still ranks as one of the finest of its kind in the country.

The Education Department reaches over 16,000,000 men, women and children each year with its museum programs and extension services. Since its establishment in 1884 it has touched the lives of hundreds of millions of people. Its programs are not intended to be static, rather do they change steadily with changes in our culture and with the ever-progressing pattern of general education.

In the realm of direct teaching, thousands of school children participate each week in the Museum's all-day program for elementary school classes. These classes leave their schools in the morning and visit the Museum from 10:00 A.M. to 2:00 P.M. Their day is one of carefully planned experiences under the direction of a trained corps of Museum teachers. Classes for visually and physically handicapped children and from the Junior and Senior High Schools of

New York weekly attend the Museum to participate in programs designed to meet their needs.

It is impossible to measure the influence which these Museum visits have upon these children and young people. Thousands of letters have been received from children who have come to the Museum programs. Teachers and parents have in turn written to express their appreciation of the Museum's teaching services. Sometimes the children are asked to write a short composition as a result of their Museum experience.

One of these compositions, full of misspelled words and poor grammatical construction represents, however, a remarkable achievement. It was the first composition ever written by a ten-year-old boy, a member of a special class of children with retarded mental development which had been given a Museum lesson on harmful insects. According to his teacher, the child never before had been sufficiently stimulated by anything he had heard or seen, in or out of school, to want to write about it. The following paragraph we have reproduced in the exact punctuation, spelling and spacing of the original.

> "the teacher also taught us about the flyhe said
> that fly is not our friend he also told us that
> each mail and feamail fly had a 120 children
> andthere Children children had A120 and so on
> and he said that it wouldn'ttake a week before
> the whole citu had flys and that these flys
> had hooks on theref feet and when a fly lights
> on our food and eats a little he also leavs some
> thing that that he leavs is germs and that we
> could stop them by not Ieaving food around on
> the table the way that the fly gets the dirt

and dease he may fly on something he flys on the
gabich that people throaw away he flies on menoue
and all sorts of things that are not clean
the teacherdid not say very much about the cocroch
but i do know this much they are no friend to this woild"

The American Museum's educational program provides material for persons of all ages. Pre-school age children meet weekly to attend the meetings of the Story Hour Club. The toddlers in this group ranging from 3½ to 5 years of age are the youngest participants in the Museum's activities. There are programs for students in the nearby colleges and universities, technical schools and schools of fine and applied arts. Ever since 1929 a series of special evening lectures and film programs has been given for adults belonging to the Evening Elementary School Students Association. Most of the students in this group are foreign-born and the museum program is designed to help these adults become oriented to their new homeland.

The Museum's policy has been and still is to meet the needs of the members of the community it serves. These needs may be highly specialized. A group of missionary students may ask to be briefed on the natural history of Burma or some other far away land to which they will soon be sent. A group of student nurses may want instruction in human physiology and comparative anatomy; a group from a trade-union may ask for information on the background of their particular craft; a group of officers of the United States Armed Forces needs specialized briefing on the natural history of the Arctic, or a garden club may wish to become more fully acquainted with the insect pests that ravage their flower gardens.

MAN AND NATURE

For many years the Museum has made wide use of motion pictures. Many of these were "shot" on Museum expeditions. Others have been acquired by gift and purchase and form the stock-in-trade of an excellent circulating loan collection of films which are sent all over the United States and succeed in bringing the Museum to millions who may never have the opportunity of entering its buildings. These films are also shown at the Museum on Wednesdays and Saturdays and attract, entertain and instruct many thousands of visitors.

There is also a vast loan collection of lantern slides and kodachromes. These pictures cover the entire range of the science of natural history. Thousands of them were taken on Museum expeditions and in this manner, the results of what may have been primarily a scientific expedition become of tremendous educational value.

A large collection of actual specimens and of well-constructed dioramas or miniature groups is made available to schools and similar institutions. The Museum maintains a fleet of delivery trucks which daily transport this and other material to the hundreds of schools in the City of New York.

The American Museum has pioneered in work with the visually and physically handicapped. As early as 1909, a room was set aside for the adult blind where specimens were accompanied by labels written in Braille, and since that time, many programs have been planned and executed for both adults and children with visual or physical handicaps.

Ever since Professor Bickmore began his lectures for school teachers, the Museum has been active in the field of teacher training. After all, it is one of the pioneers in visual education and has been teaching by this method even before it was introduced into the public schools of New York. Courses such as "The Mechanics of Visual Instruction," "Uses of Museum Materials in Elementary Education," "Natural Science and Conservation," "Diorama Making and Craft Techniques for Classroom Use," and "Descriptive Astronomy for Teachers," are but examples of the many courses which the Museum has presented to teachers. These courses are accredited by local colleges and by the Board of Superintendents of the New York City Public Schools. Members of the scientific departments continue the traditional cooperation of the Museum with local universities such as Columbia and New York University and serve on the faculty of these institutions in addition to fulfilling their museum responsibilities.

The Museum has made full use of the press, radio and more lately, television to bring its treasures before the public eye. It began its radio career in 1924 when it presented a series of talks on natural history for children over Station WEAF. Since that date thousands of broadcasts, many over nation-wide hookups, have carried the story of Man and Nature to untold millions of people. The Museum cooperated with the television industry while the latter was still in the experimental stages, and since the advent of actual telecasting many programs on the Museum and its contents have been produced. One of the earliest was telecast from the Museum's exhibition halls and featured the dancing and singing of a full-blooded Onondaga Indian who was a member of the Museum's staff.

A very popular contemporary program is "Around the World with Dance and Song." In it brilliant artists of other countries and cultures combine education with exotic entertainment. It has brought to the Museum auditorium stage such outstanding exponents of ethnic dance as Pearl Primus, Jean Leon Destine, Tom Two-Arrows, Hadassah, La Meri, Uday Shankar and Asadata Dafore.

Perhaps the largest crowds to participate in the educational programs are those coming to the Annual Children's Book Fair. Over one hundred thousand children and adults have visited this Fair in a four-day period. The greatest single-day's attendance was before World War I when over fifty thousand came in a single day to see a special exhibit on Tuberculosis which led to the establishment of the Museum's Department of Public Health.

Every year is exciting for the American Museum of Natural History. Its continuous growth means continuous new exhibitions, new researches, new expeditions and new educational programs. Each new year brings even more visitors and additional contacts through its publications and the press, radio and television.

The vast building, a small city in itself with its almost six hundred employees, its 23 acres of floor space, 15 acres of which are devoted to public exhibition, presents a tremendous problem in housekeeping alone. A veritable sea of glass encloses the thousands of exhibits and the chore of indoor window washing is alone a Herculean task. To maintain the buildings, to provide for the acquisition of specimens, research, preparation, exhibition and educational activities cost almost two and one quarter million dollars in 1950. The City of New York provided almost one million dollars of this annual cost. The rest was provided by income from the investment of its endowment funds, gifts, income from memberships, and sales and services.

THE WORLD OF NATURAL HISTORY

The Yankee Stadium in New York City is a busy place. The sports events which draw big crowds to this Stadium often fill its seats with over 60,000 spectators at a single event. In 1950 the Museum's attendance of almost 2,500,000 far exceeded that of the Yankee Stadium for the same year.

The dream of young Bickmore has been fulfilled. The story of the world of nature and of man which the American Museum of Natural History has studied, and portrayed in word and exhibit, is the story of life itself, life as it was, as it is, and as it will be.

Like the very organisms whose evolution the Museum's studies and exhibits portray, the institution itself has evolved and is still evolving. The Museum today is a far cry from the simple beginnings in the old Arsenal Building in Central Park. The Museum after another century may look back on its present state in almost the way we now look back to the Museum of 80 years ago. The dream of young Albert Smith Bickmore was broad enough and adaptable enough to allow the growth and change that has led to the present vast and amazingly productive complex known as the American Museum of Natural History. This idea and its fulfillment by the thousands of men and women who provided the funds and the labor to bring the story of Man and Nature to people everywhere is a saga of a truly cooperative effort of many men and women for the good of all.

INDEX

Adirondack Forest Group, 151
Adler Planetarium in Chicago, 33
Administration, 240, 281
African Ethnology Hall, 224-227
African mammals, 151-162
Agassiz, Louis, 14, 16, 21, 27, 28, 110, 146, 147, 242
Agassiz Museum at Cambridge, 14
Age and Area hypothesis, 258
Age of Mammals, Hall of the, 82
Age of Man, Hall of the, 82, 88, 174, 177, 184
Ahnighito (meteorite), 29-31
Ainu of Japan, 222
Akeley, Carl, 46, 47, 151-162, 283, 285
Akeley Memorial Hall of African Mammals, 151-162
Albert National Park, 162
Allen, Dr. J. A., 147, 250
Allen Hall of North American Mammals, 164
Allosaurus, 87
Amadon, Dr. Dean, 268
Amateur Astronomers Association, 44
American Cancer Society, 292
American Museum of Natural History, origin of, 14
American Ornithologists' Union, 249
American Society of Ichthyologists and Herpetologists, 111
Amphibians, 119-126
Amphibians and Reptiles, Hall of, 119
Amphibians and Reptiles, Department of, 120, 121
Amphibians, fossil, 80, 82
Anatomy, human and comparative, 170
Andrews, Dr. Roy Chapman, 79, 89, 90, 145, 147, 256, 263-267
Andros Island, Bahamas, 101
Animal Behavior, Department of, 106, 274, 293
Anning, Mary, 84
Antelopes, African, 158-162

Antelope, Indian, 163
Anthony, Dr. Harold E., 154, 157, 277, 293
Anthropology, Department of, 168, 169, 189, 239, 273, 274, 289, 293
Ants, 106
Ants, army, 274, 275
Archaeological expeditions, 189, 254-259
Archaeology, Hall of Mexican and Central American, 191-197
Archaeopterix (fossil bird), 133
Aristophanes, 247
Armstrong, John C., 99
Aronson, Dr. Lester, 274, 277, 293
Arsenal building, 19, 20, 22, 98, 146, 312
Asiatic Natives, Hall of, 219-224
Astronomy, 29
Astronomy, Department of, 31
Astronomy, early exhibit, 32
Attendance, 297
Atwood, Dr. Wallace, 33
Auduboniania, 131
Audubon, John James, 131, 134
Auk, great, 26, 268
Australia, 232
Axelrod, Alex, 169
Aztec, 190-196
Aztec calendar stone, 194, 195

Bailey Collection (minerals), 52
Baker, George Fisher, 54
Baluchitherium, 92
Bandelier, Dr. Adolph, 189
Barnum, P. T., 16, 26, 150
Barracuda, 102
Barro Colorado Island, 248, 275
Barton, Otis, 118
Basket Makers, 211
Bataan, 268
Bathysphere, 118
Bauersfeld, Dr. Walter, 32, 35
Beach, Dr. Frank, 274
Bear, Florida black, 165
Bear Mountain nature trails, 306
Bears, Alaska brown, 165

INDEX

Bears, grizzly, 164
Beck, Rollo, 140
Beebe, Dr. William, 118, 277
Bee hive, 105
Beetles, lady, 103
Belgium Congo, 162
Bement, Clarence S., 52-54
Bement Collection (minerals), 52, 53
Benin bronzes (African) 226, 227
Bennett, Dr. Wendell, 169
Benson, John P., 166
Berkey, Dr. Charles, 89, 267
Bickmore, Albert Smith, 14, 16-20, 24, 26, 83, 98, 146, 147, 169, 305, 309, 312
Biogeochemistry, 241
Bird, Junius, 169, 190, 198, 273
Birds, 130-145, 247-252, 268-273
Birds and Man exhibit, 133
Birds, Department of, 130, 144, 147, 267
Birds, local, 131, 144, 250
Birds of Paradise, 139-141
Birds of World Hall, 141
Birds, Sanford Hall of the Biology of, 131-133, 135, 272
"Birthday Cake," Emperor's, 59
Blakeley, Dudley, 157
Blind, work with, 309
Bliss, George, 17
Blodgett, William T., 17, 21
Blood, Captain Neptune, 269
Boas, Dr. Franz, 169, 189
Bogert, Charles, 127-129, 277, 293
Bogoras, Waldemar, 169, 189, 219, 220
Bones, preparation of, 283
Bongo Group, 159
Book Fair, Children's, 311
Boston Society of Natural History, 16, 21
Bowfin, 117
Brandreth, C., 143
Breder, Dr. Charles M., 111, 277
Brewster, William, 249
British Museum, 15, 19, 83, 148
Bromfield, Louis, 298
Brontosaurus, 85, 87, 288
Bronze Age, 177
Brown, Dr. Barnum, 71, 79, 261
Brown, James, 17
Bruce, Matilda, 54
Bryozoa Group, 99, 100
Buffalo (bison), 164, 187, 207, 210, 211
Buffalo, African, 158
Buffalo, Indian use of, 210
Buffalo, water, 164
Building stones, 62
Bumpus, Dr. Herman C., 264
Burden, Mrs. Douglas, 119

Bush Negroes (Dutch Guiana), 225
Butler, Albert, 157
Butler, Howard Russell, 32, 46
Butterflies, 103, 106
Byrd, Admiral Richard E., 277

Cadwalader, J. L., 143
Calendar stone, Aztec, 40
California, University of, 254
Calkins, Dr. Gary N., 80, 244
Camels, fossil, 93
Camera, Akeley, 155, 156
Cancer research (Sloan-Kettering Institute), 292
Cannibalism, 235
Cape, feathered Hawaiian war, 231
Carcharodon (giant fossil shark), 74, 114
Cardano, 139
Carter, J. C., 143
Carter, T. Donald, 293
Castillo cave, 255
Caucasoid race, 188
Cave Man, 178, 181
Cave models, 64, 65
Cazier, Dr. Mont A., 104, 277, 293
Celestial spheres, 33
Cement gun, Akeley, 156
Central American archaeology, 191-197
Central Asiatic expeditions, 79, 88-91, 255-257, 262-267
Central Park, 15, 17, 19, 21, 22, 144, 146, 312
Cetaceans, 145, 146, 149, 166, 264, 265
Chalcedony figurine, 58
Chaney, Ralph, 89
Chapin, Dr. James, 268
Chapman, Dr. Frank M., 130, 131, 144, 147, 247-252, 264, 268
Cheirolepis (fossil fish), 171, 172
Chicago Museum of Natural History, 88, 153
Chicago Academy of Sciences Museum, 33
Childs, Dr. George, 289
Chimpanzee, 162, 173
Chinese, 188, 221, 222
Choate, Joseph H., 18
Chubb, S. Harmstead, 288, 289
Civil War, 248
Clam, giant, 100
Clams, man-trap (Tridacna), 103
Clark, Dr. James L., 154, 157, 264, 283
Cobb's Island Bird Group, 250
Colbert, Dr. Edwin H., 80, 93
Coles, Robert R., 34
Colgate, Robert, 17
Collections, circulating, 309
Collections, study, 289, 290

INDEX

Colorado Museum of Natural History, 184
Columbia University, 73, 79, 80, 89, 243-245, 265, 310
Columbia University Biological Series, 244
Conchology, Department of, 98
Conservation, 241
Constable Collection (shells), 98
Cook, Captain James, 136
Cooper, James Fenimore, 252
Cooper, Theodore, 170
Cope Collection (fossils), 72, 77, 98
Cope, Edward Drinker, 77, 78, 243
"Copeia," 111
Copper Age, 177
"Copper Man" (Chile), 201
Copper Queen Mine model, 64
Coral, 100-102
Coral Reef Group, 100-102
Cortez, Hernando, 191-194
Coues, Elliot, 249
Crampton, Dr. Henry E., 80, 244
Cretaceous Hall, 82
Cro-Magnon Man, 182
Crooke Collection (shells), 98
Curator, definition of, 241
Curran, Dr. C. Howard, 292
Cynognathus (fossil reptile), 172
Cypress Swamp Group, 125

Dafoe, Dr. Allan, 145
Dafore, Asadata, 311
Dana, Charles A., 18
Darwin, Charles, 246
Darwin Hall, 97, 104, 246, 247
Daulte, Frank, 98
Davidson, Jo, 136
Davis, Leonard M., 46
Davison, F. Trubee, 19
Dawson, Charles, 181
Day, Dr. John C., 98
Dean, Dr. Bashford, 73, 74, 110, 244
De Cordoba, Francisco, 192
Deer, 163, 164
DeLong, Mrs. Edith Haggin, 53, 56
Destine, Jean Leon, 311
Deutches Museum, Munich, 33
Devil fish (manta), 116
Devonian, 74
Diamonds, 56
Diaz, Bernal, 193
Dickerson, Mary C., 121, 239
Dinosaur eggs, 89-91, 267
Dinosaurs, 77, 83-91
Dinosaurs, paintings of, 88
Dinosaurs, first North American, 84
Diplodocus (dinosaur), 85
Diplovertebron (fossil amphibian), 82, 172
Dixon, Roland, 169
Dodge, William E., 16, 21

Dogs, wild, 163
Doyle, Robert, 269
Drummond Jade Collection, 58-60
Drums, African, 227
Dubois, Dr. Eugene, 178, 236

Earthquakes, 65
East African Plains Group, 160
Easter Island stone head, 231
Eastern Woodlands Indians crafts, 215-217
Eastman-Pomeroy African expedition, 262
Eclipse expedition, 45
Ecology, 298, 302
Economic Geology and Petrology, Hall of, 65
Education, adult, 308
Education, Department of, 239, 240, 305, 306
Educational program, 305
Edwards Collection (insects), 105
Ekhorn, Dr. Gordon, 169, 191, 273
Elasmobranchs, 116
El Cano, 139
Elephants, 150, 153-155, 158, 163, 183
Elkhorn Ranch Group, 152
Elliot Collection (birds), 130
Elliot, Dr. Daniel G., 20, 130, 153, 276
Ellis, Dr. Brooks, 291
Emerald, engraved, 57
Eohippus ("dawn horse"), 93, 94
Eolithic (Dawn Stone Age), 176
Eryops (fossil amphibian), 82, 121
Eskimo Hall, 202-204
Ethnic dance program, 310, 311
Eusthenopteron (fossil fish), 171
Evolution, Osborn's laws of, 245
Exhibits, planning, 282
Expedition, first official, 276
Expeditions, 260-279, 284
Experimental Biology, Department of, 274
Exploration, "Golden Age" of, 262
Extension services, 309

Face (from fish to man), 170
Falkenbach, Otto, 79
Farrand, Livingston, 169, 189
Faunce, Wayne M., 240
Faunthorpe, Colonel J. C., 163
Field, Benjamin H., 17
Figurine, chalcedony, 58
Finance, 311
Fire-walking, Tahitian, 232
Fish, armored (Dinichthys), 74, 75
Fish, coral reef, 102
Fish, deep-sea, 118
Fish, deep-sea angler, 109
Fish, giant bulldog, 75

315

INDEX

Fish, Great Barrier Reef, 113
Fish, modern (Teleosts), 74, 76
Fisher, Dr. A. K., 249
Fisher, Dr. George Clyde, 32, 45, 46
Fishes, Age of, 73
Fishes and Aquatic Biology, Department of, 99, 110, 111, 239
Fishes, food, 117
Fishes, game, 112, 113
Fishes, Ganoid, 76
Fishes, Hall of, 76, 109, 111-118
Fishes, Hall of Fossil, 74-76
Fishes, lobe-finned, 76
Flaherty, Robert, 202
Fly, model of, 106
Foliage, preparation of, 286
Folsom Man, 187, 188
Foraminifera, fossil, 291
Ford, Dr. James, 169, 273
Forestry and general botany, 298
Fossils, 71-95
Fossil aquarium, 75
Fossil Fish Hall, 73-76
Fossil teeth, studies of, 243
Fossil trees, 69
Fossils, preparation of, 288
Foulke, W. Parker, 84
Franklin Institute, Philadelphia, 34
Frogs, 121-126
Fuertes, Louis Agassiz, 143, 250, 251

Garber, Dr. A. Z., 89
Ganoid fish, 117
Garpike, 117
Gems, 53-60
Geology, 51
Geology, Department of, 52, 99
Geology, physical, 60-69
Geronimo, 212
Gertsch, Dr. Willis, 293
Ghost Dance shirts, 208
Giant Panda Group, 166
Giant Sable Antelope Group, 161
Gibbon, 163, 173
Gila Monster, 124, 125
Gilliard, E. T., 263, 268-274
Glacial Period, 174
Glass blower, 97, 98
Gloria maris (rarest shell), 100
Gobi Desert, Mongolia, 79, 89, 256, 260, 262-267
Goddard, Dr. Pliny, 169, 255
Gold, Costa Rican (Mayan), 192
Goodwin, George, 293
Gordon, Dr. Myron F., 292
Gorilla Group, 156, 162, 173
Gowanus Canal, Brooklyn, 145
Grand Canyon of the Colorado, 63, 64, 164
Granger, Dr. Walter, 79, 89, 91, 259, 267
Grant, President Ulysses S., 25

Gratacap, L. P., 54
Green, Andrew H., 17
Gregory, Dr. William K., 80, 110, **170**, 171, 173, 244-246, 292
Grey, Lord, 142
Grey, Zane, 112
Griffith Park Planetarium, 34
Gudger, Dr. Eugene, 115
Guerry, Joseph, 157
Guyot, Professor, 242

Haas, Dr. Otto H., 69
Habitat groups, 166, 167, 283-288
Hadassah, 311
Hadrosaurus (dinosaur), 84
Haida canoe, 204
Haines Collection (shells), 98
Haines, William H., 17
Hall Collection (fossils), 27, 62, **72**
Hall, Professor James, 15, 27
Harriman Park, 164
Harty, M. F., 240
Hastings, Margaret, 269
Hay, Clarence, 169, 191, 273
Hayden, Charles, 34, 49
Heidelberg Man, 180
Henry, Professor Joseph, 26
Hermann, Adam, 79
Herpetology, Department of, 99, 110
Herrick, Professor C. J., 243
Hitchcock, Professor C. H., 62
Hittell, Charles, 143
Hoffman, Malvina, 158
Holder, Dr. Joseph B., 98, 146
Hollins, H. B., 143
Homo sapiens, 177
Hope, John, 154, 157
Hopi snake dance, 127-129, **214**
Horse Hall, 288
Horses, fossil, 93, 94
Horsfall, Bruce, 143
House fly model, 247
Hrdlicka, Ales, 170
Huanchaco, Peru, 46
Hughes, Charles Evans, 252
Hunt, George, 169
Huntington, C. P., 170
Hurricanes, 41-43
Hutchinson, G. Evelyn, 241
Huxley, T. H., 78, 242
Hyde, B. Talbot, 170, 189
Hyman, Dr. Libbie, 239

Ice Age, 175
Ichthyology, Department of, 99, **110**
Ichthyosaurus, 84
Iguana, rhinoceros, **124**
Iguanodon, 84
Incan agriculture, 199
Incan metalcraft, 200
Incan pottery, 199, 200
Incan textiles, 199

316

INDEX

Incas of Peru, 198-201
Indians, American, 188, 190
Indians, Hall of the Eastern Woodlands, 214-218
Indians, Jesup Hall of North Pacific Coast, 204-207
Indians, North American, 202-218
Indians of Manhattan Island, 215, 216
Indians of the Southwest, Hall of, 211-214
Indians, Plains, 190
Indians of South America, Hall of, 197-202
Indians, South American, 201
Indonesia, 232
Insect Hall, 103-107
Insects, importance of, 107, 108
Insects and Spiders, Department of, 104, 105, 289, 292, 293
Insects, backyard, 104
Insects, live, 289
International Game Fish Association, 111, 113
Invertebrate Department, 97, 99, 110
Invertebrate Paleontology Exhibit, 69, 70
Invertebrate Zoology, Department of, 99
Ipiutak culture (Eskimo), 189
Iron Age, 177
Iroquois Indians, 216, 217
Iselin, Adrian, 17

Jacobsen, Philip, 169
Jacques, Francis L., 103
Jade, 58, 59
Jaguars, Temple of (Chitzen-Itza), 195
Jansson, Arthur, 157
Japan, 265, 266
Japan, Crown Prince of, 94
Japanese Collection, 222
Jasper National Park (Canada), 164
Java Ape-Man, 178
Jesup, Morris K., 17, 19, 62, 72, 78, 169, 189, 204, 219, 229, 245, 250, 262
Jesup, Mrs. Morris K., 143
Jesup North Pacific Expedition, 262
Jochelson, Waldemar, 169, 189, 219, 220
"Jumbo," the elephant, 150
Junior Astronomy Club, 44-46
Jurassic Hall 81

Kidd, Captain, 192
Kien Lung, Emperor, 59
King, William, 177
Klauber, L. M., 128, 129
Knight, Charles R., 38, 88, 95, 182

Kokoda Track, New Guinea, 268
Koryaks (Siberia), 220, 221
Krakatoa eruption, 68, 69
Kroeber, Dr. A. L., 190

Lacandon Forest, Mexico, 260, 275
La Meri, 311
Lamont, William, 269
La Monte, Francesca, 239
Lang, Charles, 79
Lang Ivory Collection, 225, 226
Lanier, Charles, 58
Laufer, Berthold, 169, 221
LeClear, Bonnie Wallace, 54
Lecture, first, 305
Leigh, Jonas, 157
Lerner, Michael, 111-113
Lerner Research Station at Bimini, 113, 292
Lewis, Mrs. Isobel M., 45
Library, museum, 239, 303, 304
Lindbergh, Charles A., 252, 268
Lion, African, 159
Lion, Asiatic, 162
Living Invertebrates, Department of, 101
Lizards, 119, 120, 122-125
Lizards, Dragon, 119, 120
London, University of, 242
Long, Professor Roger, 33
Loubat, Duke of, 170
Low, Seth, 243
Lowie, Robert, 169
Lucas, Frederick A., 152
Lumholtz, Carl, 170
Lungfish, 116, 117
Lutz, Dr. Frank E., 104

MacMillan, Donald, 204, 277
Magellan, 139
Maintenance, 280-282
Malay, 235, 236
Mammal collection, 146
Mammals, Age of, 175, 176
Mammals, African, 151-162
Mammals and Birds, Department of, 146, 147
Mammals, Department of, 130, 145-147, 154
Mammals, fossil, 91, 95
Mammals, Hall of Biology of, 149-151
Mammals, New York State, 151
Mammals, North American, 151
Mammals, North Asiatic, 166
Mammals, sea, 147, 165, 166
Mammals, South Asiatic, 151
Mammoths, 183
Man and Nature Halls, 299
Man, Age of, 175, 176
Man, prehistoric, 168-186
Manhattan Square, 22-25

317

INDEX

Maori (New Zealand), 229-231
Marine Zoology, Department of, 98, 99, 146, 147
Marlin Group, 112, 113
Marsh, Othniel Charles, 77-79
Mantell, Dr. Gideon, 83, 84
Mason, G. Frederick, 157
Mastodonts, 183
Matthew, Dr. William Diller, 72, 79
Maximilian of Neuwied, Prince, 20, 121, 130
Maya, 190-193, 195, 275
Mayr, Dr. Ernst, 140, 246, 268
McGregor, Dr. J. H., 80, 179, 244
Mead, Dr. Margaret, 169, 228, 239, 273, 293
Mearns, Dr. E. A., 130
Meganthropus paleojavanicus (Great man from Stone Age Java), 236
Melanesia, 232-235
Mesozoic Era, 76-91
Messina, Angelina, 291
Metal, Ages of, 177
Meteorological exhibit (Willett's Memorial), 41
Meteors, 30, 31
Meteorites, 29-32
Metropolitan Museum of Art, 19, 21, 22, 74
Mexican archaeology, 191-197
Mexico, Valley of, 191-197
Micronesia, 232
Micropaleontology, 97
Micropaleontology, Department of, 291
Microscopic creatures, 96, 97, 99, 100
Miller, W. DeWitt, 147
Milwaukee Public Museum, 153
Miner, Dr. Roy Waldo, 100
Mineralogy, 51
Mineralogy, Department of, 54
Mongolia, 266, 267
Mongoloid race, 187-190, 203, 219-224
Mongols, 256
Montezuma, 193, 194
Moon, trip to the, 48
Moose Group, Alaska, 164, 165
Morgan, J. Pierpont, 17, 21, 53, 54, 266
Morgan, Sir Henry, 192
Morton, Levi P., 17
Mosasaur (fossil lizard), 84
Mosquito model, 104, 247
Motion pictures, 309
Mountain lion, 164
Mountain sheep, 164
Mueller, Herman, 97, 98, 289
Mummies (Peruvian) 198-201
Murphy, Dr. Robert Cushman, 130, 268, 277

Museum, contributions of, 294
Museums, International Committee on, 297

Nanook of the North, 202
Narahara, V. 157, 289
National Geological Survey of China, 179
Nature trails, 306
Natural History Magazine, 93, 304
Natural History Magazine, Junior, 304
Natural History of Man, Hall of the, 170
Neanderthal Man, 177, 178, 181, 182
Negroid race, 188
Neilson, Ernest, 289
Nelson, Dr. Nels C., 89, 169, 252-259
Newell, Dr. Norman D., 69
New Forest Group (birds), 141, 142
New Guinea (expedition for birds), 268-273
New Guinea Group (birds), 139
New Mexico, University of, 213
New Stone Age, 176, 182, 185
New York Academy of Sciences, 16
New York Botanical Gardens, 19
New York City, relations with, 18
New York Mineralogical Club, 55
New York University, 310
New York Zoological Park, 19
Nichols, Hobart, 143
Nichols, John T., 111
Noble, Dr. G. Kingsley, 274
North American Bird Groups, 142-144
North American Mammal Hall, 151, 164, 165
North Pacific Coast Indian tribes, 205, 206
Notharctus (fossil primate), 173

Ocean Life, Hall of, 97, 99, 100, 145, 147, 149, 151, 166
Oceans, 118, 136
Octopus, 103
Okapi Group, 161
Old Stone Age, 176, 181, 182, 185, 187, 189
Olmec head, giant stone, 197
Olsen, George, 90
Opossum, 172
Ornithology, pioneers of American, 248, 249
Orthodontists, American Board of, 292
Oryx, 161
Osborn, Professor Henry Fairfield, 19, 72, 78, 79, 88, 92, 94, 95, 241-247, 252, 255, 266
Osborn, William Henry, 241

INDEX

Osborn, Virginia Reed Sturgis, 241
Osten-Sacken Collection (insects), 105
Osteological laboratory, 283
Ostracoderms (jawless fishes), 74
Owen, Sir Richard, 15, 83
Oxford University, 15

Pachycephalosaurus (dinosaur), 88
Pacific Islands, 228-236
Paddlefish, 117
Painting, background, 287
Paleontology, Department of, 288
Palisades Interstate Park, 306
Parícutin (volcano), 67, 68
Parish, Henry, 18
Parr, Dr. Albert Eide, 240, 297, 299
Pearl Divers Group, 100, 102, 103
Pearl oyster (Meleagrina), 102, 103
Peary, Admiral Robert E., 29-31, 204, 277
Peking Man, 179-181
Pelée eruption, 66, 67
Pembroke, College, 33
Pepper, George, 257
Peru (Incas), 197-201
Peruvian Guano Group (birds), 137
Petersen, George, 157
Phelps-Dodge, A. G., 17
Phelps, I. N., 17
Philadelphia Academy of Natural Sciences, 16, 21, 84
Philippine Hall, 228, 229
Phytosaur (Fort Lee, N. J.), 83
Pierce, H. C., 143
Pigeons, passenger, 133-135
Piltdown, Man, 181
Pine Plains, N. Y., 298-303
Pirates, 192
Pithecanthropus erectus (Java Ape-Man, 178, 236
Plains Indians, Hall of the, 207-211
Plains Indians tribes, 208
Planetarium, Copernican, 32
Planetarium dome, construction of, 46-48, 156
Planetariums, early, 33
Planetarium, Hayden, 23, 30, 32, 34-39
Planets, 39
Plesiosaurus (fossil lizard), 84
Pole, South, 277
Pole stars, 36, 37
Polynesia, 232, 235
Pomeroy, Daniel E., 153, 262
Poor, H. W., 143
Pope, Clifford, 89
Porpoises, 149
Potlatch ceremony, 205
Potter, Howard, 17
Pough, Dr. Frederick, 67
Pough, Richard H., 241

Preparation Department, 264, 282-286
Precessional motion, 37
Prehistoric animals, 71-95
Preserved heads, Maori, 230, 231
Primate Hall, 151
Primates, fossil, 95
Primus, Pearl, 311
Princeton University, 241, 242
Protoceratops andrewsi, 90, 91
Protozoa, 96
Pteranodon, 83
Pterodactyls, 83
Publications, popular, 304
Public Health, Department of, 311
Public Instruction, Department of, 294
Pueble Bonito, 189
Pueblo, Hopi, 213
Putnam, Dr. F. W., 169

Quetzal, 195

Raddatz, R., 157
Radio, 310
Ramsey, Dr. Grace F., 306
Rancho La Brea tar-pits, 81, 183
Rattlesnake, 127, 195
Rays, 116, 117
Reeds, Dr. Chester, 176
Reiss, Dr. Bernard F., 293
Relief models (geological), 63
Reptiles, 119-130
Reptiles, Age of, 76-91
Reptiles (fossil flying), 83
Research, aeronautics, 292
Research programs, 291-293
Rhinoceros, African, 162
Rhinoceros, Asiatic, 163
Richardson, Jenness, 276
Roberts, Marshall O., 17
Robinson, Coleman, 98, 105
Robley, Major General G., 229
Rock shelters, Indian, 215
Rockwell, Robert, 154, 157
Roosevelt Medal, 251
Roosevelt Memorial, 17, 131, 151, 158, 246, 247
Roosevelt, President Franklin Delano, 17
Roosevelt, President Theodore, 16, 142, 152, 158
Roosevelt, Theodore, 16, 17, 21
Rothschild, Lord, 131, 140
Rotifer Group, 100
Rubies, 57
Russell, Dr. Charles, 295, 297

Sacrificial Stone (Aztec), 195
Sailfish Group, 112
St. Gaudens, Augustus, 136
St. Pierre, destruction of, 66
Salamanders, 121, 122

INDEX

Salmon, Eleanor, 291
Sanford, Dr. Leonard, 131, 140, 272
Sapphires, 57
Saville, Marshall, 169
Scenes, behind the, 280-294
Schermerhorn, F. A., 143
Scherer, Fred, 289
Schlosser, Dr. Max, 179
Schneirla, Dr. T. C., 106, 274-276, 293
School relations, 305, 306
Schuyler, Mrs. Philip, 143
Scientific names, 86
Scientific staff, 238-259
Scott, Professor William B. 78, 242
Scutosaurus, 82
Sea, Glory of the, (Conus gloriamaris), 100
Sea lions, 166
Seals, 166
Seismograph, 65
Senff, Charles H., 229
Seymouria (fossil reptile), 82, 172
Shankar, Uday, 311
Shapiro, Dr. Harry, 168, 228, 235, 273, 277, 292, 293
Shark, blue, 116
Shark, giant fossil, 74, 114
Shark Group, Mako, 112
Shark, hammerhead, 116
Shark, man-eater or white, 114, 115
Sharks, spiny, 75
Shark, thresher, 116
Shark, whale, 113, 114
Shell collections, 98
Sherman, Benjamin B., 18
Sherwood, Dr. George H., 157
Ship-followers Group (birds), 137
Shrunken heads (Jivaro Indians), 202
Siberian natives, 219-221
Silurian, 74
Simpson, Dr. George G., 79, 246, 277
Sinanthropus (Peking Man), 179-181
Siple, Dr. Paul, 277
Skates, 116
Sloane, Eric, 41
Sloth, giant ground, 184
Smith, H. H., 276
Smith, H. I., 169
Smith, Dr. Hugh M., 229
Smithsonian Institution, 16, 21, 26, 110, 130
Snakes, 122, 123, 125-130
Solomon Island Group (birds), 138, 234
South Pacific Hall, Natives of, 229-236
Southwest Indian crafts, 212, 213
Southwest Indian tribes, 212
Soviet culture, study of, 293
Spang Collection (minerals), 52

Spalding, Mr. and Mrs. Keith, 112
Sperry Gyroscope Company, 292
Spiders, 104
Spiers, Dr. Francis, 242
Spinden, Dr. Herbert, 169
Stanford University, 253
Star of India (star-sapphire), 55
Stars, 43
Star, Sixteen-rayed sea, 103
Stefansson Vilhjalmur, 204, 277
Stegosaurus (dinosaur), 86, 87
Stephens, D. Owen, 46
Steward Collection (shells), 98
Steward, D. Jackson, 17
Stewart, Alexander T., 17
Stone Age Culture, Hall of, 177
Stone Ages, 176-186
Storer, Albert, 98
Storer Collection (shells), 98
Strömgren, Dr. Elis, 37
Strong, Oliver S., 244
Struthiomimus (dinosaur), 88
Stuart, Robert L. 17, 19, 26
Stuart, Mrs. Robert L., 130
Sturgeon, shovel-nosed, 117
Sullivan, Louis, 169
Sun Dance, 209
Sunday openings, 249
Sunfish, ocean, 112
Sun, Hall of the, 39
Svenson, Dr. Henry, 298
Swan, Lucille, 180
Swanton, John, 169
Swiss Lake Dwellers, 186
Swordfish, life history of, 113

Teacher training, 309, 310
Teaching, direct, 306
Tahitian crafts, 231, 232
Tannery, 283, 285
Tate, Dr. George H. H., 293
Tattooing, Maori, 229-231
"Tax Collector" (Peruvian), 197, 198
Taxidermy, Akeley method of, 285, 286
Taxidermy, Department of, 147, 264, 282-286
Taylor, Will S., 204, 205
Teeth, "Dragon's," 89, 179
Teeth, evolution of, 292
Teit, James, 169
Teleost fish, 117
Telescope-making class, 44
Television, 310
Templeton-Crocker expedition, 231
Tenochtitlan (Mexico City), 191-194
Terry, James, 169
Thermoregulation in reptiles, 293
Thompson, William Boyce, 54
Thomson, Albert, 79
Tibetan collection, 222-224
Tierra del Fueguans, 201

320

INDEX

Tiffany Collection (gems), 53
Tiger Group, Siberian, 166
Tigers, 163
Tiger, saber-tooth, 183
Time capsule, 40
Titanotheres (fossils), 91, 92
Tizoc (Aztec ruler), 196
Toads, 123, 124
Toad, Surinam, 123
Tongareva (Penrhyn Island), 102
Tonnelier, Georges, 58
Topaz, 57, 58
Topaz crystal, giant, 58
Tortoise, giant fossil, 83
Trachodon (dinosaur), 88
Travois, Indian, 208
Tree-ring dating, method of, 214
Trepanation (Incan), 200, 201
Trevor, J. B., 143
Triceratops (dinosaur), 88
Tridacna (giant clam), 100
Tring Collection (birds), 131, 140
Tropical air castle, Chapman's, 248
Tschopik, Dr. Harry, 169, 273
Tuna Group, 112, 113
Turtles, 122, 124, 125
Tusk, largest fossil elephant, 184
Two-Arrows, Tom, 311
Tweed, William ("Boss"), 18
Tyrannosaurus rex (dinosaur), 87

United States Fish Commission, 110, 136
United States National Museum, 78, 148
United States Naval Observatory, 45
United States Navy, 265
Upper Nile River Group, 160

Vaillant, Dr. George, 169, 191
Vaux, Calvert, 24, 25
Vedray Collection (animals), 20, 21, 130
Vernay, Arthur S., 163
Vernay-Faunthorpe Hall of South Asiatic Mammals, 151, 162, 163
Verreaux Collection (animals), 17, 20, 21, 130
Vertebrate Paleontology, Department of, 72, 77-80, 241, 243, 245, 288
Villard, Henry, 170
Volcanoes, 65-69

Von Koenigswald, Dr. G. H. R., 179, 180, 236

Walrus, 166
Warburg, Felix M., 298
Warburg Memorial Hall, 298-302
Ward's Natural History Establishment, 85, 152, 153
Warren mastodon, 183
Warren, William R., 170
Water Hole Group, 158
Weber, Rudolf, 228
Weidenreich, Dr. Franz, 180
Weitzner, Bella, 169, 273
Weyer's Cave, Virginia, 65
Whale, baby sperm, 145
Whale, blue, 149, 150, 264
Whales, 145, 146, 149, 166, 264, 265
Whaling murals, 166
Wheeler, Dr. William Morton, 99
White, Alexander, 19
Whitfield, Professor R. P., 27, 52, 146
Whitney Building, 131
Whitney family, 131
Whitney, Harry Payne, 131, 136
Whitney, Mrs. Harry Payne, 140
Whitney Memorial Hall of Pacific Bird Life, 135-140, 234
Whitney South Sea expedition, 140
Whitney, William C., 136
Wilson, Alexander, 133
Wilson, E. B., 244
Wilson, James Perry, 46
Winthrop, Mrs. Robert, 143
Wissler, Dr. Clark, 169, 190, 257, 258
Wolfe, Catherine L. 98
Wolfe, John David, 17, 19, 98
Wolves, timber, 165
Woodward, Smith, 181
World War II, 262, 268, 292
Wortman, Dr. Jacob, 79, 94

Yakut (Siberia), 221
Yale University, 274
Yankee Stadium, 312
Yellowstone Park, 164
Yosemite Park, 164

Zeiss projection planetarium, **32-37**
Zimmer, Dr. John, 268
Zodiac, 40
Zoology Department, 146

England – p. 141
Switzerland – p. 141
Siberia – p. 166
Gobi Desert – p. 85
Peking, China – p. 179
Burma – p. 56
Philippines –
India – p. 162
Abyssinia – p. 158
Upper Nile – p. 160
Belgian Congo – p. 152
Komodo Island – p. 119

An Index Map of
The World of Natural History

Expeditions of the American Museum of Natural History have explored almost every part of the world. You may read about some of their work in the countries named above on the pages indicated.